THE HISTORY OF BLOOD TRANSFUSION IN SUB-SAHARAN AFRICA

PERSPECTIVES ON GLOBAL HEALTH

Series editor: James L. A. Webb, Jr.

This series publishes innovative studies that draw upon perspectives from the natural sciences and social sciences to shed light on important issues in global public health. The books in Perspectives on Global Health interest students and practitioners and are appropriate for adoption in undergraduate and graduate courses in global public health.

The History of Blood Transfusion in Sub-Saharan Africa, by William H. Schneider

Global Health in Africa: Historical Perspectives on Disease Control, edited by Tamara Giles-Vernick and James L. A. Webb, Jr.

THE HISTORY OF BLOOD TRANSFUSION IN SUB-SAHARAN AFRICA

William H. Schneider

Ohio University Press
Athens

Ohio University Press, Athens, Ohio 45701
ohioswallow.com

Printed in the United States of America
Ohio University Press books are printed on acid-free paper ∞ ™

23 22 21 20 19 18 17 16 15 14 13   5 4 3 2 1

Library of Congress Cataloging-in-Publication Data

Schneider, William H. (William Howard), [date] author.
The history of blood transfusion in Sub-Saharan Africa / William H. Schneider.
pages cm — (Perspectives on global health)
Includes bibliographical references and index.
ISBN 978-0-8214-2037-9 (pb : alk. paper) — ISBN 978-0-8214-4453-5
1. Blood—Transfusion—Africa—History. 2. Blood banks—Risk management—Africa.
3. AIDS (Disease)—Epidemiology. I. Title. II. Series: Perspectives on global health.
RM171.S34 2013
362.1784096—dc23

2013026688

CONTENTS

ACKNOWLEDGMENTS

A book as broad in scope as this has only been possible thanks to extraordinary support and assistance from many people and organizations. This includes a grant from the Fulbright-Hays Faculty Research Abroad Fellowship Program (no. P019A050018), a visiting researcher award at the Brocher Foundation (Geneva), sabbatical leave from the School of Liberal Arts at Indiana University–Purdue University Indianapolis, and support from the Baker-Ort Chair in International Healthcare Philanthropy at the Indiana University School of Philanthropy at IUPUI.

Among the personnel at blood banks and transfusion services who opened their records and agreed to be interviewed were Jack Nyamongo (Kenya National Blood Transfusion Service), O.W. Mwanda (Nairobi University Medical School), Peter Kataaha, Paul Senyonga, and John Watson-Williams (Uganda Blood Transfusion Services), Esau Nzaro and Aisu Steven (Mulago Hospital Blood Bank), Lamine Diakhaté and Saliou Diop (Centre national de transfusion sanguine, Senegal), Dora Mbanya (Centre hospitalier et universitaire, University of Yaoundé Faculty of Medicine), Juhani Leikola (Finnish Red Cross), Evelyn von Steffen (formerly of the International Federation of Red Cross and Red Crescent Societies, Geneva), and Myriam Malengreau (Université catholique de Louvain).

There are numerous archives and libraries around the world whose collections were invaluable for this book. They were graciously made available to me by many people with a great deal of patience. They include Helen Pugh and Emily Oldfield (British Red Cross Museum and Archives), Dirk Schoonbaert (Institute of Tropical Medicine Antwerp), Luc de Munck (Flemish section of the Belgian Red Cross), Pierre Dandoy (Archives africaines, Service public fédéral Affaires étrangères, Brussels), Stéphane Kraxner (Institut Pasteur, Paris), Aline Pueyo (Institute of Tropical Medicine, Marseille), Margrit Schenker (Swiss Red Cross), Grant Mitchell (IFRC), Sophie Chapuis (International Museum of the Red Cross and Red Crescent, Geneva), Tomas Allen (WHO Library, Geneva), and Maria G. N. Musoke (Makerere University Medical Library, Kampala).

Several colleagues have offered encouragement as well as patience as this project was brought to a conclusion. They include Patrick Aeberhard, Ernie

Drucker, Bob Einterz, Ellen Einterz, Tamara Giles-Vernick, Didier Gondola, Holly Hanson, Guillaume Lachenal, Preston Marx, and Cees Smit-Sibinga. Their help is largely responsible for both the instigation and relevance of this work. Any errors or questionable judgments are entirely my own.

Staff and research assistants at IUPUI—Jennifer Broome, Kelly Gascoine, Judi Izuka-Campbell, and Jennifer Smedley—contributed time and energy well beyond normal expectations of job descriptions. Jim Webb and Gillian Berchowitz at Ohio University Press, plus anonymous readers' comments, have shown an understanding and appreciation that have been quite gratifying.

Finally, but foremost, it is a pleasure to publicly thank my wife, Laurie, for her patience and "adventures" involved in bringing this book to a conclusion.

ABBREVIATIONS

BRC	British Red Cross
CNTS	Centre national de transfusion sanguine (Senegal)
ERCS	Ethiopian Red Cross Society
Fomulac	Fondation médicale de l'Université de Louvain en Afrique Centrale
IFRC	International Federation of Red Cross and Red Crescent Societies
LRCS	League of Red Cross Societies (after 1983 League of Red Cross and Red Crescent Society; after 1991 became IFRC)
PEPFAR	President's Emergency Program for AIDS Relief (United States)
UMHK	Union minière du Haut Katanga (Belgian Congo)
WHO	World Health Organization

INTRODUCTION

The history of blood transfusion is a subject of interest to historians of medicine and scholars of Africa, as well as those who work in and follow developments in global health. Like today, the introduction of new medical practices in Africa at the beginning of colonial rule held great promise for generally improving health. In the case of transfusion, as this study shows, there were advances to be sure, but also some setbacks. Most important, blood transfusion could not be put in place quickly or exactly as it was practiced in the countries of Europe or in other developed countries. The procedure required some important modifications in order to be widely practiced in Africa; hence, its history offers many lessons for those currently interested in economic development and global health.

Blood Transfusion and the Origin of AIDS

This book grew out of the response to a question from colleagues studying the origin of AIDS. As is often the case with unexpected epidemics, the most immediate concern was about the cause of the disease after it was identified in 1981. Once that cause was discovered a few years later, an extensive search began for the origin of the virus. Despite speculation about such things as conspiracies, fallout from germ warfare research, and divine retribution, by the late 1980s virologists identified simian immunodeficiency viruses of chimpanzees and monkeys in Central and West Africa as the antecedents of the human immunodeficiency viruses that caused AIDS.[1]

One important feature of simian viruses, however, complicated the question of the origin of AIDS. According to scientists, the simian viruses from which the human immunodeficiency viruses (HIV-1 and HIV-2) were adapted have existed in African chimpanzees and monkeys for thousands of years. Moreover, humans have long had contact with these animals as sources of food and pets, but a person can tolerate relatively well a simian virus infection from cuts or bites.[2] So the question was, why did the adapted human immunodeficiency viruses become pandemic only in the last thirty years?[3]

Blood transfusion was not the first explanation suggested; rather it was that a chance, random mutation produced the particularly deadly HIV virus. This theory was reinforced by early evidence that suggested the human virus was new. An extensive search for human tissue and serum found that the earliest sample infected with HIV antibodies was from 1959.[4] Using this sample and multiple other later ones, researchers have taken advantage of the fact that HIVs mutate at a predictable rate so that they can compare the changes in DNA sequences of the HIV samples and estimate approximate dates of their common ancestor virus. Accordingly, they have estimated that the oldest human immunodeficiency virus dates from the period between 1921 and 1933.[5]

Even if this human virus is of relatively recent origin, the problem with the random-mutation explanation is that a dozen other human immunodeficiency viruses have been discovered that have also adapted from simian ones and from regions as far away as a thousand miles.[6] This has led many researchers to seek another explanation of what changed to make it possible for the simian viruses to begin adapting into human ones at different locations in Africa in the first half of the twentieth century.

The most obvious source of change in much of sub-Saharan Africa at the end of the nineteenth and beginning of the twentieth century was the establishment of European colonial rule. Some provocative hypotheses have been offered to explain the specific developments that could have increased the opportunities for adaptation of viruses as well as their epidemic spread. For example, demographic and social change, or altered patterns of environmental exploitation could have increased passage of viruses from simians to humans or between humans.[7] These suggestions have usually been found in concluding sections of articles on the scientific basis of dating the emergence of viruses, although two recent book-length treatments of the origin of AIDS have summarized, and in one case attempted to analyze, them.[8] These are broad generalizations requiring detailed research to be verified. In addition, these explanations have usually been suggested by scientists, only a few of whom have actually ventured into archives, but none with historical training in the critical evaluation of sources.[9] Therefore more serious scholarship by historians is essential before drawing any conclusions.

My colleagues' theory was that the multiple adaptations were facilitated by new, albeit unintentional means of transmitting disease between humans in Africa in the last century. Their research focused initially on the use of insufficiently sterilized needles in mass campaigns against disease that began during colonial rule, in order to show that this might have been a new way for someone infected with a simian virus to transmit it to another person. They maintained that this so-called serial passage increased the chance for the adaptation of new human immunodeficiency viruses.[10] These results raised the question of whether another otherwise lifesaving procedure, blood transfusion, began early and frequently enough in Africa also to have facilitated adaptation of SIVs to new HIVs and their epidemic spread. In this case even less frequent use of transfusion might have had a greater impact because of its unparalleled efficiency of transmitting disease, nearly 100 percent. To their credit, they recognized their limitations as historians, and they sought additional expertise.

Africa and the History of Transfusion

My colleagues' ignorance of the history of blood transfusion in Africa was not just the result of their lack of historical training. In fact, there has been very little written on Africa even though the history of blood transfusion has been told many times in many ways. Its role as a reanimating force—still the term used by French transfusion services—is perhaps the earliest characterization, dating back to the sixteenth century, and it continued into the nineteenth century, for example in an 1804 English caricature of Napoleon receiving injections of courage from the blood of a tiger.[11] At the beginning of the twentieth century, transfusion was described as a "miracle of resurrection" by George Crile's wife, who assisted at one of his first successful transfusions.[12]

Most histories of transfusion have emphasized technical discoveries. A classic case is the first chapter of Geoffrey Keynes's *Blood Transfusion* (1922),[13] which describes the attempts beginning with Richard Lower in England and Jean-Baptiste Denys in France in the 1660s, following William Harvey's discovery of blood circulation. Keynes also mentions the efforts of James Blundell in the first part of the nineteenth century, but the main purpose of the book is to demonstrate the much more successful techniques discovered after 1900, which were first used on a large scale during the Great War (1914–18), in which Keynes (whose older brother, John Maynard Keynes, was the famous economist) had recently served. Subsequent histories, usually in the form of articles or introductory chapters of transfusion texts, have also focused on new techniques developed during the rest of the twentieth century.[14] Those histories freely skipped around time periods and locations to document the discoveries and innovations of those who built on their predecessors around the world to perfect the increasingly important medical procedure.

This book differs from most of those histories in its focus on one geographical area. Only recently have there been such histories—for example Susan

Lederer's *Flesh and Blood* (2008), which examines the history of both organ transplantation and blood transfusion in the United States.[15] That book illustrates one reason for focusing on a specific geographical area: it permits greater detail about how a medical procedure such as transfusion is affected by and has a broad impact on society. This might be true for other medical procedures, yet for blood transfusion it is especially interesting because the main therapeutic "medicine" is blood from another person. The impact of transfusion is dramatic and lifesaving.

In addition, a geographical focus makes it possible to identify features of the practice that are peculiar to that setting, hence it is useful for comparative study. The conditions in sub-Saharan Africa were so different from those in the United States or Europe that one can easily see the value of identifying what was similar and what was different in its history. For example, the fact that there was no analogous treatment in traditional African medicine meant that the introduction of blood transfusion offers an excellent case of how modern biomedicine is adopted or modified in settings outside modern societies, a question of great importance in global health today. Sub-Saharan Africa, for purposes of this study, includes all of Africa south of the Sahara, with the exception of South Africa, where the very different political, social, and demographic circumstances of white rule, especially after 1900, produced a health system that was very different from the rest of sub-Saharan Africa. For example, a 1975 editorial in the *South African Medical Journal* reported that 86 percent of blood in South Africa was donated by whites, while at least 60 percent of patients were estimated to be black.[16]

Approaching the History of Transfusion in Africa

The research for this book is an attempt to help answer the question about the role of transfusion in the origin of the AIDS epidemic by showing in detail when, where, and how blood transfusion was introduced to Africa.[17] Because there is very little scholarly study of the subject, another goal of this book is to describe the major changes in the introduction and expansion of blood transfusion from the first reports of the practice, during colonial rule in the 1920s, to independence and the appearance of AIDS, at the end of the twentieth century. But blood transfusion is not a remote, detached medical procedure. On the contrary, it requires people to have blood taken from them, as well as patients willing to have the blood of others placed in their bodies. As a result, the history of blood transfusion reveals a number of features of African societies that adopted this novel medical treatment, including who donated blood, the uses of transfusion, the question of risk from contaminated blood, and the extent to which transfusion might have spread HIV and other diseases.

The sources used for this history include the records of most of the colonial powers in Africa—Britain, France, and Belgium—as well as several European Red Cross societies and the International Federation of Red Cross and Red Crescent Societies (IFRC). Major medical libraries and archives around the world also provided published and unpublished reports of the African health services, and four African countries provided surprisingly good records of hospital and transfusion services. A number of former health officers who worked in Africa were of great assistance in answering questions and occasionally providing more detailed histories of their service.

There are limits worth noting in a book of this scope. First is the difficulty in providing enough depth of analysis for an area so large and diverse as sub-Saharan Africa. Even limiting the study to the British, French, and Belgian colonies, it is impossible to give the depth of treatment that all colonies warrant. In an effort to resolve this problem, Kenya, Uganda, Senegal, and the Congo/Zaire[18] have been chosen for more extensive examination because of their size, geographical variety, and availability of records. Hopefully subsequent studies will follow on these and other histories. A second problem is the lack of records about the Africans who made up the donors and patients. This book does not pretend to be a social history of transfusion, but it is hoped that indirect evidence from numbers and observations will provide enough detail to represent not only the subjects of transfusion but also in this case the source of the "medicine" (blood) for treatment. Finally, and related to the previous point, this subject risks amplifying the divide and stereotypical differences between Western and traditional medicine. To be sure transfusion was not a practice with any analog in traditional African medicine, but that does not mean that opposition or incompatibility were inevitable. In fact, a major finding is that transfusion was adopted quickly and with relatively little resistance. Moreover, there were definite differences in the practice in Africa compared to Europe and elsewhere, such as securing donors and uses of transfusion, which means that this was not a simple process of adoption of Western medicine.

The essential requirements for transfusions to begin in Africa were doctors trained in the techniques, donors, and patients in need of and willing to receive transfusions. The first reports in Africa were in the early 1920s, and organized transfusion practices had been developed before the Second World War. The records between the two world wars show not only that all conditions existed in sub-Saharan Africa that were necessary for blood transfusions; they also suggest that the numbers were limited primarily by the availability of Western medical doctors and facilities to do transfusions. There is also an indication of how innovation took place, usually through connections to people and resources outside the usual colonial medical structures, for example, the Red Cross, missionaries, universities, and mining companies.

Transfusion became a regular part of modern medical treatment in sub-Saharan Africa from 1945 to independence. The means by which this occurred differed significantly according to the colonial ruler. For example, in francophone Africa the government attempted to implement a policy of centralized blood collection in Dakar, Senegal, to supply blood to all colonies in French West Africa. In the British and Belgian colonies local initiatives and the Red Cross were much more important in creating transfusion services. In Uganda, this practice led to the Red Cross expanding the number of collection centers from the capital, Kampala, to other regions of the protectorate. The common underlying reason for growth everywhere was the Africans' acceptance of donating and receiving blood. Equally important were the increased expenditures in colonies on health, particularly new hospital construction, because transfusions were done in the hospital setting. In addition, new and simpler techniques developed during the Second World War made transfusion easier to practice in sub-Saharan Africa.

Following independence, in the 1960s, transfusion continued to grow in Africa and the organization of services entered a new phase. Most newly independent countries accelerated expansion by building provincial and district hospitals to serve regional and local needs. These hospitals usually had the ability to do transfusions, but with only a few exceptions governments left it to the local hospitals to arrange for their own blood collection, sometimes with the assistance of the Red Cross and unpaid donors, sometimes with a paid service, and sometimes both. Thus, there was a general swing away from centralization and its high costs, toward a mixed organization with at best limited regional services, but also hospital-based means to supplement or complement the collection, testing, and distribution of blood for transfusion. Hospitals thus developed a number of options for blood collection, all of which were driven by an increase in the use of transfusion for medical care and the corresponding need for more donors.

The final phase in the development of African transfusion services began after the economic crisis of the mid-1970s, when African countries were unable to provide resources to continue, let alone keep up with, new techniques in transfusion medicine. This constraint limited their ability to draw and store blood or extend transfusion to more remote regions. Problems were exacerbated by growing concerns with testing and safety, such as the need to screen for hepatitis B, a new disease that was discovered well before the appearance of HIV. One response was to seek funding from developed countries, especially in Europe, North America, and in Japan. When successful, the result was a recentralization of transfusion services because donor countries found that it was more efficient and safe and gave them better ability to monitor how funds were used. For example, foreign assistance in Burundi and Rwanda followed this pattern, as did Ethiopia, but not all countries were able to secure outside funding. Other pressure for centralization came from the growth of programs at the World Health Organization and the International Red Cross, both of which helped secure funding and coordinated offers of technical assistance for setting standards of blood safety beginning in the mid-1970s. They also co-sponsored the first African blood transfusion workshops, beginning in Burundi in 1976 and the Ivory Coast in 1977. Thus, when the AIDS crisis hit Africa, less than a decade later, mechanisms to support and coordinate efforts to monitor and insure a safe blood supply were already in place.

Three crucial features of blood transfusion in Africa are particularly noteworthy compared to elsewhere: what transfusions were used for, who donated blood, and safety. One surprising finding is that reluctance of patients to receive the blood of others offered relatively little impediment to the adoption of transfusion in Africa. To be sure, as elsewhere, there were fears and myths that arose and for the most part were allayed by practitioners. Moreover, the needs were so extensive, and the successful results of transfusion were so dramatic, that if anything, the overuse of transfusion became a bigger problem than resistance on the part of patients. Hospital records are scarce in the early period, but after the 1980s statistics show that transfusions were done increasingly for maternity and pediatric (anemia) use. Other evidence of the uses of transfusion includes posters from the Red Cross archives, which, in an effort to encourage Africans to donate blood, prominently featured how the blood would be used.

Another surprising conclusion is that sufficient donors were generally found in Africa so that the blood supply was able to meet the growing use of transfusion. The best explanation for this success in meeting the need for blood donation was the flexibility of these practices. Most notably, hospitals were innovative in how they adapted to their circumstances in order to secure blood donors. This pragmatism ran counter to some expectations of resistance and irrational opposition by Africans. One way to interpret this was that the "medicine" given to Africans in transfusion, blood, was possessed in the same amount and with the same control by Africans as anywhere else in the world. There were few or no drug companies or expensive chemical manufacturing or rare materials that had to be purchased. As far as donors were concerned, therefore, the history of blood transfusion offers a good example of Africans' ability to organize and adapt their health care well when the materials were available to them.

With some significant exceptions, the most important institutions in finding ways to obtain blood for transfusions were hospitals. Initially, donors were found as needed, from family, friends, and those willing to be on call. Later, donors were provided food and drink and in some places payment, both officially and unofficially. Most important in this process was not whether donors were "voluntary" or paid, since in Africa notions of obligation and compensation were more complicated than in Europe or North America. Rather it was the size and facilities of hospitals or collection centers that dictated whether blood was drawn for immediate use, as with smaller hospitals, or for storage and distribution to other hospitals, as with larger facilities.

The main finding about the history of risk from blood transfusion in Africa is that those giving transfusions were well aware of risk from the start, and that is not surprising given the disease environment. That does not mean there was much that could be done to prevent most disease transmission through transfusion. For example, malaria is not screened to this day in countries where it is endemic (most of sub-Saharan Africa). If potential blood donors with malaria were screened, the wide prevalence of the disease would mean drastically reducing the number of transfusions, and those transfused would face exposure to the disease in any case.

The blood transfusion practices in sub-Saharan Africa were incapable of detecting HIV, as was the case initially with even the most sophisticated screening in more developed health care systems. In contrast, Africans did not have the resources to improve immediately their ability to detect and screen for the virus once it was recognized. As a result, given the wide use of transfusion, it was unfortunately an important reason for the initial spread of AIDS. Yet African health officials were not without means to understand and respond to the new danger, thanks to forty years of experience and a framework of appreciating long-standing health risks. Both by screening high-risk donors and taking advantage of outside support for testing for HIV, as well as technical advice and training, the reduction of AIDS transmission through contaminated blood transfusion was one of the quickest and most successful responses to the epidemic in sub-Saharan Africa.

1 BLOOD TRANSFUSION BEFORE THE SECOND WORLD WAR

Blood transfusion is one of the most important lifesaving discoveries of modern scientific medicine. The first effective procedures were demonstrated only at the beginning of the twentieth century, although attempts in its present form began shortly after William Harvey's discovery of blood circulation, in the seventeenth century. Long before that, most cultures had recognized the significance of blood, which played a prominent role in many rituals and customs of healing.[1]

The history of how blood transfusion was introduced to Africa is important beyond its dramatic role in saving lives. As a well-defined medical procedure that was unprecedented in traditional society, transfusion provides a clear case of how an innovation of modern technology, relatively new to Western medicine itself, was introduced to Africa. As a result, it offers an example of Africans' responses to Western medicine, as well as Westerners' views of Africans and appropriate medicine for them. In addition, although blood transfusion was similar to other new medical technology such as x-rays or anesthesia, it is much richer in its human and social complications because of the symbolic and cultural significance attributed to blood.

Another revealing feature of transfusion is that it required a weighing of risks and benefits for patients, in this case the potential for the dramatic saving of life compared to the risk of dangers such as disease transmission. Transfusion thus offers an indication of how well these subtle and complicated judgments were made in the African setting. This latter point has become dramatically more important since the 1980s, following the emergence of new infectious diseases such as AIDS. Transfusion, an obvious lifesaving procedure, also opened a new means that had never existed before of transmitting pathogens between humans with unprecedented efficiency. Historian of medicine Mirko Grmek recognized early in the AIDS epidemic (1989) that blood transfusion

> was rapidly developed [after the First World War] and became one of the most effective and frequently used therapeutic methods. Still, it was

only toward the middle of the century that the practice of transfusion opened a gap in the barrier which, from an epidemiologic point of view, separates the blood of one human being from that of others. . . . For the microorganisms [certain viruses], blood transfusion had formerly been a narrow path, used only in exceptional occasions for transmission of some sporadic infections. Today it has become the royal road requiring delicate and difficult monitoring.[2]

Evidence of the First Transfusions

The first issue of the *Annales de la Société belge de la médecine tropicale* (1921) contained a report by Belgian doctor Émile Lejeune that described a patient he treated on the East African front at the end of the First World War. Lejeune had gone to the Belgian Congo after graduating as a young doctor from the University of Louvain in 1911 and very quickly gained notice by establishing one of the first services that went regularly out to villages to test, treat, and inoculate large numbers of Africans. In fact, it became the model for the much more famous and widely implemented campaigns of Eugène Jamot, beginning in Cameroon in the 1920s.[3] When the First World War broke out, Lejeune and other doctors in the Congo were mobilized and went to the East African front, where the British were engaged in what turned out to be a four-year campaign against a German-led force in Tanganyika. The mobilization of Africans and Europeans produced a far greater number of casualties from disease than combat, and Lejeune helped staff the hospitals that were established to care for the sick and wounded.[4]

Among his patients was a European officer of colonial troops who suffered from hemoglobinuria (blackwater fever),[5] and Lejeune prescribed a standard course of cure: "treatment by all normal measures: physiologic serum, injections of hypertonic saline solution, adrenaline, Murphy sugar [drip]." After a few days without improvement Lejeune was discouraged. "General conditions are frankly becoming bad; the patient is very weak, delirious; the pulse, despite medication to stimulate it, is hardly perceptible and very accelerated."

Lejeune consulted his colleague, Dr. Giovanni Trolli, who was also there on temporary duty from the Belgian Congo Medical Service, and they concluded that the situation was "desperate." Lejeune therefore decided "to attempt a blood transfusion as the ultimate therapeutic trial." To be sure, the transfusion he gave was a measure of last resort, but the patient's fever broke the next day, and although his blood count took longer to return to normal, he was discharged and returned to Europe in two months.

Thus was reported one of the earliest records of a blood transfusion in sub-Saharan Africa (excluding South Africa and Rhodesia). Whether it was in fact the first transfusion (likely not), this case reveals a number of things about the beginning of blood transfusion there. First, it was surprisingly early. It occurred in 1918, only fifteen years after the reports of the first successful transfusions in Europe and America that ushered in the era of modern use of a long-dreamed of therapy: giving the blood of one human to another. In fact, it was not until after the First World War that techniques were worked out that made the procedure viable for widespread use in Europe and America.[6]

Second, Lejeune showed that transfusion was feasible in the African setting. Admittedly, he was a European-trained doctor who treated a European patient, using blood drawn from another European. This reflected one of the main reasons noted by scholars for the beginning of tropical medicine: care for colonial interests, European settlers, administrative officials, and Africans employed in mining and other colonial enterprises. But with the exception of settler colonies, such as Kenya and Rhodesia, the number of Europeans in Africa was quite small. In the Congo Free State of Leopold II, there were around five hundred Europeans in 1901, and an estimated twenty-five to thirty doctors, mostly in the towns of Boma and Léopoldville. This number grew to fifty-nine doctors in 1910, not counting those employed by missionaries and private companies.[7] So, to provide for the minimum health needs of relatively few Europeans, there was obviously a capacity to care for some Africans, even if that paled in comparison to what their numbers and disease burden required.[8]

By the end of the nineteenth century, the most important setting for the practice of modern biomedicine was a hospital. So where Lejeune did his transfusion was similar to others established by the new colonial governments in sub-Saharan Africa, usually in ports, colonial capitals, or areas of important economic activity, where significant numbers of Europeans were likely to be living. These hospitals were numerous enough in some parts of Africa by the First World War to establish many Western medical practices in the African colonies. There were government hospitals in British East Africa—in Entebbe, Mombasa, and Nairobi—within ten years of colonial rule, and by 1919 the Germans had established hospitals in Tabora, plus the Ocean Road and Sewa Haji facilities in Dar es Salaam.[9] Although their benefits eventually reached the African populations, more or less, how much is not the point here. Rather it is that by 1918 the level of medical infrastructure and knowledge in sub-Saharan Africa was sufficient to practice blood transfusion.

To illustrate this point, consider that Lejeune used 500 cc of citrated blood, "from a healthy, solid European, with no apparent defects," but he did not match blood types. Although test sera and a method to determine compatibility were greatly improved by war's end such that results could be done in minutes, it is not surprising that Lejeune was unaware of this, given his very remote location.[10] But even without test sera, there were procedures to guard against incompatibility that involved giving a small amount of blood to the patient at first, to see if there was an adverse reaction, before transfusing the remainder. This was crude but effective and justifiable if the life of a patient was at stake.[11] In fact, this is exactly what Lejeune did in his 1918 transfusion. He reported giving an "anti-anaphylactic injection" of 5 cc of blood from the donor and waiting five minutes to see of there was a reaction before transfusing the rest of the blood. So Lejeune was aware of blood groups, their importance in transfusion, and a way to match blood, in effect *in vivo*.

In concluding remarks, obviously aimed at other doctors practicing in Africa, Lejeune pointed out the implications of his success: "All doctors in the Congo have seen patients die in this manner [complications from hemoglobinuria]. The successful results we obtained could be repeated in other cases of this type. . . . Transfusion is an operation that can be done anywhere. Blood was drawn by means of strong needles on a Dieulafoy syringe. . . . Transfusion is nothing more that an intravenous injection. I did it slowly (a quarter of an hour)."[12]

It is impossible to say with certainty when the first blood transfusion was made in Africa. It is a very large continent, the number of Western doctors was small, and the records of practice were quite varied. Reports published by doctors such as Lejeune were among the important sources describing the beginning of blood transfusion. Because it was an unusual practice at first, transfusion was likely deemed noteworthy by both doctors and medical journals that published accounts of their experience. Informative as these accounts may have been, they certainly did not record every transfusion, and in any case, once the novelty soon wore off, they would be of less interest. For example, neither Lejeune nor Trolli published again on transfusion, but it is likely that they practiced what Lejeune advocated in his 1921 article when they returned to the Congo in the 1920s. Both, in fact, served there for an extended period, with Trolli becoming chief medical officer of the colony in 1925 and Lejeune remaining in the Congo as a private physician in northern Katanga after finishing his government service.[13]

Other important sources of information were the routine medical services reports that became routine in African colonies by the 1920s. The standard forms for surgical interventions did not mention transfusion, however, until much later, when the procedure was more widely used. Generally, it was only after transfusion services were established by hospitals, usually beginning in the late 1940s, that medical reports provide a continuous record of transfusions.

There is nonetheless indirect evidence of transfusions before that time, such as the accounts of laboratory tests found in the annual medical reports. Hospital laboratories kept good records of examinations and analyses that had become part of hospital routines, and many included blood group tests in their annual reports, often well before the surgical reports began recording transfusions used in an operation or treatment. These blood group determinations could only have been for transfusion purposes, since paternity and forensic testing were almost unheard of in Africa at this time. In fact, these exceptions were duly noted when they were occasionally performed.[14] Laboratory reports, therefore, provided a sustained, if indirect, record of blood transfusion beginning between the wars.

Starting with the example of Lejeune, plus the direct and indirect evidence in published articles and unpublished reports, it appears that the first transfusions were done in Africa at the beginning of the 1920s. This means that the conditions necessary for a transfusion—a patient in need, a donor, and a doctor with knowledge and means—existed at a relatively early time, even compared to the first transfusions in the world in the modern era, which began only in the decade before the First World War. The record also shows, however, that only after the Second World War were transfusions done widely and in large numbers in Africa. Understanding this timing requires further examination of questions about the setting in which the introduction of this lifesaving Western medical technique took place.

Blood Donation and the Uses of Transfusion in Africa

There were three broad conditions required for transfusions to take place in Africa or anywhere else: patients in need, a source of blood, and facilities with someone knowledgeable of a method to transfuse the blood. The first two of these conditions depended on Africans' attitudes toward Western medicine, and more specifically whether they understood the need for transfusion and were willing to do something as unusual as donate and receive blood. Surprisingly, these were not the biggest hindrances or limiting influences on when blood transfusion began. To be sure, there was a natural reluctance to do something as unusual as allow blood to be taken from or introduced into one's body; and there are numerous anecdotal examples and a few systematic studies of resistance to, as well as persistent racist observations about, the inability of Africans to understand Western medicine. For example, Meghan Vaughan has described the resistance of Africans in northern Nyasaland to smallpox eradication efforts in the 1930s. In addition to compulsory infliction of some pain, the measures also entailed, "the curtailment of movement, the segregation of villages, the banning of funerals and the burning of victims' huts."[15]

These examples were exceptions to the general and relatively quick acceptance by Africans when Western medicine became available. There was not only little resistance but Africans also eagerly responded when its effectiveness was demonstrated. An early example was the rapid success of the Anglican doctor Albert Cook, who came to Uganda in 1897. In fact, his fellow missionaries already there were afraid the Africans' acceptance would distract them from religious conversion. Shortly thereafter a Church Missionary Society dispensary in the colony attracted over two hundred patients a day within months of opening.[16]

A later example was the speed, surprising to some Western observers, with which Africans generally accepted injections and other Western medicines. This was frequently noted by outsiders who feared overuse by patients demanding injections or medicines, no matter the condition. One medical officer who served in Uganda beginning in the late 1930s reported cases where blood donors thought that the act of donating blood had the curative power of an injection because a needle was used.[17]

Blood transfusion required even more explanation, especially for donors, but it was one of the Western medical techniques whose value was immediately and obviously demonstrable, in Africa as elsewhere. When Grace Crile, wife of the American surgeon George Crile, described the results of his first transfusion on a human, which she assisted as a nurse in 1906, she recalled, "I stood at the foot of the operating table and witnessed the miracle of resurrection."[18] Thus, in large measure, because it worked so well, transfusion became a part of modern medicine throughout the Western world soon after a safe way was found to transfer the blood from donor to patient, at the beginning of the twentieth century. The experience of the First World War helped resolve some initial problems, and in the 1920s transfusion shifted from wartime use for injuries sustained in battle to its more common civilian uses to replace blood loss from various accidents and diseases, as well as in childbirth. All these conditions existed in Africa, as well as another endemic to the region, severe anemia.

Many expressed doubt that Africans would allow their blood to be taken or subject themselves to such a radical procedure as introducing the blood of another into their bodies. For example, early reports of transfusion in the

Belgian Congo relied on recovering patients in hospitals as the source of blood, people with little power to decline.[19] Likewise, a similar approach was used to persuade African troops in Kenya to donate blood during the Second World War, but there was so much resistance that a special study was done to learn why.[20] In Senegal, one of the early practitioners of transfusion, Gaston Ouary, expressed strong doubts about Africans donating blood for fear of becoming weak from blood loss or somehow contracting the disease of the patient receiving blood.[21]

These fears and occasional reluctance to donate blood proved not, however, to be the obstacle that some Western observers feared. In the end, the obvious benefit that a transfusion produced was coupled with adaptation and persuasion to obtain the necessary blood donors. Writing in 1960, at a critical juncture on the eve of independence in many African countries. H. C. Trowell, a British physician at Mulago Hospital in Kampala, Uganda, quickly dismissed the potential problem of finding blood donors. In "Transfusion," a section in his *Non-infective Disease in Africa,* he stated, "It is not proposed to discuss the social prejudices against blood transfusion in Africa, as within a few years these are usually overcome, and then it is usually the shortage of staff and apparatus, rather than the shortage of donors, which is the limiting factor."[22]

In fact, the overall pattern was not so different from that in the West, where a variety of methods and motivations, from patriotism to payment, have been used to secure adequate blood for transfusion. Yet according to most studies, less than 9 percent of the U.S. population (of donor age 18–65 years) donates blood in a given year. In Africa a combination of voluntary donation, appeals to obligations from family and friends, and payment have historically been used to secure an adequate blood supply.

The Development of Transfusion Technology to 1950

Of all the things that determined when and how blood transfusion came to Africa, in shortest supply were the facilities and someone knowledgeable about the procedure. Doctors were simply not available in large enough numbers in Africa to introduce blood transfusion on a wide scale until after the Second World War. The techniques they used were adaptations of those worked out in Europe and America in the first half of the twentieth century. These methods strongly influenced when, where, and how transfusion was practiced in Africa, hence it is worth reviewing them, because in the end, transfusions were given in Africa essentially as elsewhere: in hospitals, by doctors or their assistants. Thus, even more than other procedures of Western medicine, such as drug prescriptions or injections, the history of blood transfusion in Africa was linked directly to the two most important institutions of Western medicine: hospitals and doctors.

Patients and healers have long thought blood had curative and restorative power, but the effective medical use of blood transfusions is a relatively modern innovation. It was only after Harvey's discovery of the circulation of the blood, in the seventeenth century, that there was demonstrable proof of the potential benefit of transfusion, and not until the beginning of the twentieth century that effective blood transfusions entered the realm of scientifically based medical practice.

Surgeons took the lead in developing the effective techniques of blood transfusions at the beginning of the twentieth century; hence patients were treated in a hospital setting with sterile conditions, with anesthesia if necessary, and careful monitoring. These conditions were indispensable for the first effective transfer of blood from a healthy donor to a patient; in fact the initial transfusions were done by connecting the artery of a donor to the vein of a patient. This lifesaving, although long and delicate, procedure was repeated by surgeons in a number of locations who quickly added such basic refinements as measuring the amount of blood donated and preventing clotting. The discovery that sodium citrate delayed coagulation meant the end of so-called direct transfusion, where blood drawn from a donor was immediately given to the patient, usually in the same room. Now, the drawing of donors' blood into a syringe or tube could be separated from the procedure of giving it to the patient. The result immediately made transfusion easier, but the procedure remained under the supervision of a doctor. Other changes took longer to be appreciated, such as the fortuitous and simultaneous but independent discovery of blood group compatibility, which required almost a decade and more rapid testing before matching donors and patients became a routine part of transfusion.[23]

A key turning point in these new developments was the mobilization of resources and the great needs of the First World War, which offered both an opportunity and the need to refine procedures in order to give transfusions more easily and quickly.[24] These innovations, in turn, helped spread the practice in civilian medicine after the war, although transfusion still took a number of years to be widely used in medical care. Systems of obtaining donors were organized between the world wars, and as a result, the number of blood transfusions in Europe and North America grew steadily. By 1938 transfusion services in major cities (New York, London, Paris) reported five to nine thousand transfusions per year,[25] with rates of one to two hundred per hundred thousand population. This was substantial but modest as compared to more than ten times that rate after 1945.

The Second World War dramatically increased military demand, and the number of donors grew along with the development and adoption of new techniques to collect and preserve blood. After the war, these methods were rapidly introduced to meet growing demand throughout the United States and Europe, as the various collection services shifted and expanded their wartime organizations to meet civilian needs. In the Netherlands, there were 43,000 registered donors by the time the Germans invaded in 1940. Given the conditions of German occupation, that number declined during the war. After liberation and the rebuilding of health services, however, there were eighty thousand Dutch blood donors by 1953 (in a population of 10.5 million). That same year Belgium, with a population of 8.8 million, had 47,000 blood donations, while the Canadian Red Cross reported 345,000 bottles donated (in a population of just under 15 million).[26] The National Blood Transfusion Service in the United Kingdom reported over three hundred thousand transfusions annually in the immediate postwar years, a figure that climbed steadily, surpassing 1 million in 1958 (about two thousand per hundred thousand population) and reaching 1.7 million in 1972. By 1953 the United States was collecting over 4 million blood donations annually with a national transfusion rate of 2,490 per hundred thousand. The 2005 U.S. Nationwide Blood Collection and Utilization Report indicated over 15 million units collected, making the annual blood transfusion rate in the United States approximately 5,230 per hundred thousand total population or 8.6 percent of the donor age population (18–65 years).[27] Comparable figures exist for England and France.[28]

While there were many differences in the systems employed in different countries, blood transfusion services were institutionalized and became widely available in the United States, Europe, and most developed countries in the dozen years following the Second World War. This included well-organized donation, storage, and distribution methods, and testing for known contaminants. Scientific journals—such as *Transfusion,* established in 1947 by the American Association of Blood Banks, and *Vox sanguinis,* which began in 1951 and is published by the International Society of Blood Transfusion, which also began holding international congresses in the 1930s—provided means for sharing new discoveries and administrative innovations.[29]

The Knowledge and Motivation to Use Transfusion in Africa before the Second World War

The key to understanding the introduction of blood transfusion to Africa is that it was practiced by doctors in hospitals. As a result, it was the availability of hospitals and doctors with the knowledge and desire to use the procedure that was most important in determining when transfusion was used, rather than Africans' willingness to give or receive blood. There is ample evidence that these resources existed in Africa following the First World War, when the practice of Western medicine was broadly introduced to Africa.[30] By then hospitals had been built in the colonial capitals and large ports and towns. Although the extent of services varied, there was usually one chief hospital in a colony where Europeans and Africans could be treated, and often another large hospital only for Africans. These hospitals provided a base of knowledge, service, research, and training to support the expansion of Western health and medical care to the rest of the colony. Smaller towns and regional centers could subsequently develop district hospitals that varied quite widely in size and service, but each typically had at least one European doctor.

In this setting, doctors with knowledge of blood transfusion were most likely to be found in the large hospitals established in the capitals and ports at the beginning of colonial rule, and the number of doctors and hospitals increased in most colonies in the 1920s and 1930s. For example, when Trolli became head of the Belgian Congo health services in 1925, he did a census of services and found ninety-seven government doctors and thirty-six doctors attached to companies or religious orders. Then, when King Albert I and Queen Elisabeth of Belgium visited the Congo in 1928, upon their return they persuaded the Belgian parliament to create funding, including an endowment for FOREAMI (Fonds Reine Elisabeth pour l'assistance médicale aux indigènes) that was planned by Trolli and set up in 1930 for disease campaigns. By the late 1930s FOREAMI employed twenty-seven Belgian doctors, plus sanitary agents and African assistants.[31] Likewise in all of French West Africa, there were only thirty-seven doctors in 1890, but by 1910 that number had grown to 140.[32]

In British East Africa, European doctors were assisted by Indian-trained assistant surgeons, and in the French colonies by graduates of the African medical school in Dakar. An effort was made to train indigenous "dressers" in East Africa in the 1920s, but with the limited exception of Uganda, the numbers were not significant.[33] In any case, there is no evidence that these non-Europeans had the responsibility to do transfusions on their own, although they helped greatly to fill the staff of hospitals and assist European-trained doctors, who might be more inclined to do transfusions in an appropriately staffed hospital. In Dakar, the École de médecine de l'Afrique occidentale française was established in 1921 and by 1934 it graduated 148 "doctors" (although not recognized by European standards) and 191 midwives, who were drawn pretty evenly (between

40–70 each) from the colonies of Senegal, Soudan (French Sudan), French Guinea, and the Ivory Coast.[34] On the eve of the Second World War, one history of the French colonial medical service states there were 165 doctors in French West Africa, supplemented by 34 civilian doctors, 32 Russian "hygienists," and 184 African doctors from the Dakar school.[35] The hospitals were mostly government facilities, and the doctors were employed by the government as well, but in many colonies there were also hospitals of varying size and levels of care established by missionaries and other philanthropic organizations, plus hospitals created by companies in mines and plantations. The colonial governments quickly found it useful to provide subventions to retain the philanthropic institutions, because it was cheaper than replacing these facilities with government ones.

Depending on their training, the growing number of physicians in Africa came increasingly to know about blood transfusions and the techniques refined during the First World War that spread into civilian practice in Europe during the 1920s and 1930s. As the years progressed, new doctors coming to colonial posts were even more likely to know about blood transfusion from their training in British, French, and Belgian medical schools. As the example of Lejeune has shown, many of the transfusion techniques that had been simplified during the First World War were within the means of most doctors to learn and practice. If lives could be saved close to the battlefields of Europe, they could also be saved in a hospital setting in Africa. The equipment necessary included a syringe or other device to withdraw blood from a donor, plus sodium citrate to delay coagulation before the blood was introduced into the vein of the patient.[36]

Since need and sources of blood were not limiting conditions, transfusions first took place in Africa between the wars, where there were doctors trained recently enough and with the means to introduce the latest new procedures. Surveying the continent, one finds these conditions in a number of locations. First and foremost were places with sufficient European populations to warrant Western hospitals: South Africa, Rhodesia and Kenya, Mozambique and Angola, plus cities in other colonies with significant European business or government activity. In addition there were colonies with fewer Europeans, but where the metropole had invested significantly in health facilities for Africans to support economic activity (e.g., mining), or where there was sufficient development of health infrastructure to reach Africans.

One example of the knowledge of transfusion and willingness of Africans both to donate and receive blood can be found in South Africa. Although an area not included in this study, conditions were similar enough to illustrate the point early on. In a 1921 paper, J. H. H. Pirie of the South African Institute for Medical Research described testing for blood types that was inspired by the research of Ludwik and Hanna Hirszfeld during the First World War. They had done blood group tests on thousands of troops, including 250 Africans, and found striking differences in the proportions of the ABO blood types depending on country of origin.[37] What made it possible for Pirie to verify these results was his observation that "blood transfusion is a procedure which has now become so frequently employed . . . that a brief review of the preliminary tests required in order to ascertain the suitability of the donor's blood may not be out of place."[38] Pirie did not say whether he used existing blood tests of black Africans receiving transfusions, or if he tested subjects especially for his study, but his article at least demonstrates that blood transfusion was practiced routinely in 1921 at two hospitals in South Africa, where his colleagues provided him access to blood tests.

Early Transfusion Services in the Belgian Congo

There is noteworthy evidence of doctors in the Belgian Congo who followed the suggestion to repeat the successful results described in Lejeune's report of transfusion. Although, there was little European settlement in the colony, Belgian authorities made significant investments in health in the 1920s and 1930s because of business and mining interests and a government expectation of productivity benefits from healthier subjects.[39] In fact, there were reports from at least three different locations where transfusion began in this large colony before the Second World War. Although the doctors likely soon knew of each other's work, the opportunities developed independently, and there was no effort at coordination by the colonial government. Because Belgian colonial administrators took advantage of a variety of sources for medical services, the government increased the number of health facilities but hindered centralization. This same independence probably made the introduction of transfusion more likely because of multiple influences, but expansion was less likely because of limited resources.[40]

Of the three places where transfusions were reported, the one most similar to other colonies was in the capital of Léopoldville. There the Hôpital des congolais grew into a large general hospital for Africans between the wars, and although postwar plans for a new one never were achieved, additional space and updated equipment were added to serve the growing population of the capital. Unlike Kenya, Nigeria, or Uganda, where new hospitals were built with the latest facilities, including blood banks, transfusions increased at the Leopoldville hospital without much fanfare.[41] For example, as early as 1939 a doctor in the pediatric service of the Hôpital des Congolais began transfusions for severely anemic infants from a variety of causes including (malaria, worms, malnutrition, and syphilis).[42] He had previously done the procedure in Rwanda. By 1956 over seven thousand transfusions were done annually for these cases.[43] The Queen Astrid Laboratory, which serviced the hospital, reported the preparation of test sera for determining blood types as early as 1947, and by 1954 it reported doing over sixty-four hundred blood group tests.[44] The initiative for a more central blood collection and processing service started only in 1953 and from a facility in Léopoldville with outside connections: the Congo Red Cross. In fact, this outside association provoked a conflict with the Hôpital des congolais, which had its own recruitment of donors.

One of the doctors at the Hôpital des congolais who did blood transfusion after the Second World War was Joseph Lambillon, the head of the maternity service. He had first done transfusions in Africa shortly after he went to the Congo in 1938 to work in the eastern Kivu region at a hospital in Katana that was supported by the University of Louvain. Fresh from two years as an assistant in one of the top surgery services in Belgium, Lambillon was eager to introduce modern medical practices that were appropriate for the Congo. In 1940 he published an article, coauthored with the other doctor at the hospital, entitled "Étude de l'organisation d'un service de transfusions sanguines dans un centre hospitalier d'Afrique."[45] The report, in fact, referred to only thirty transfusions, but Lambillon was less interested in claiming credit for a new procedure than he was eager to demonstrate, like Lejeune before him, the viability of transfusion in the African setting. He concluded, "This note has no pretensions of innovation. But it permits us, in the end, to underscore that in the colonial setting blood transfusion is very easily done, thanks to the large number of chronic patients that are in all the native hospitals who can serve as donors. Transfusion has the very big advantage of being a striking treatment that above all is not costly, a fact which is of great importance in native medicine." Lambillon thus showed it was not lack of donors, nor Africans' rejection of the value of blood transfusion that stood in the way of using the lifesaving procedure. Doctors simply needed to use it.

The one place where Lambillon's efforts most likely had an impact was at another hospital run by the University of Louvain in Kisantu, at the other end of the colony in Lower Congo. As late as 1934, this hospital reported no transfusions, despite its unusual link back to a major medical faculty in Europe. It was in the same year Lambillon's article appeared (1940), however, that doctors at Kisantu began to treat severely anemic infants with blood transfusions.

Once they had begun, they did so in a very systematic and extensive way. Throughout the 1940s and into the next decade, Kisantu Hospital treated over six hundred infants, most under a year of age, with over twenty-two hundred transfusions annually by 1949.[46]

The most unusual and earliest report of transfusions in the Belgian Congo, however, was in yet another location with unique resources and opportunities to do blood transfusions: the medical service of the Union Minière du Haut Katanga. In 1924 doctors at the African hospitals at Panda and Elisabethville (later, Lubumbashi), in Katanga Province, published the results of studies using transfusion therapy for African workers with pneumonia. This was a cross between serum therapy and transfusion, since blood was drawn from convalescing pneumonia patients and then given to patients with active cases of pneumonia. An initial test on forty-five patients was followed by a larger study of 238.[47] The results, however, were not definitive. Although doctors recognized the risk of introducing different pneumonia strains, they concluded, "comparing our results overall, this method is the best that we have a chance to use. Compared with various other treatments and colloid therapy, . . . it represents serious progress."[48]

An even more innovative transfusion technique was reported in a 1934 note by Dr. George Valcke about an autotransfusion he practiced on an African woman in Katanga who had hemorrhaged after giving birth. He withdrew blood from her abdominal cavity, filtered it, and then reintroduced it to her as a transfusion. Valcke, who served in the Congo for over twenty years before returning to Belgium in 1933 to head the Leopold II Clinic in Antwerp, indicated he had learned the technique from Professor Joseph Sebrechts of the Catholic University of Louvain, one of the most famous Belgian surgeons, who gave a demonstration in Elisabethville in 1930.[49] Valcke's brief 1934 note responded to a lengthy article on the work of the obstetrical clinic at Kisantu Hospital, which made no reference to transfusions.[50] The doctor in charge, Antoine Duboccage, mentioned that among several cases was a severe hemorrhage ending in death. Valcke noted that Duboccage should have used the autotransfusion method. Despite this suggestion, it took a change of personnel at the Louvain hospital and the report of Lambillon's work before transfusions began in Kisantu for anemic infants.

An indication of how widespread transfusion was practiced in the Belgian Congo, and possibly other colonies in sub-Saharan Africa by the Second World War, can be found in a thesis written at the Prince Leopold Institute of Tropical Medicine in Antwerp in 1950. In it the author (listed only as L. Kok) described giving over a hundred blood transfusions in the Belgian Congo "to natives as often as possible over a dozen years." That this was not unusual is made clear in the opening sentence, which bore out the prediction of Lejeune thirty years earlier: "Today blood transfusions are done on a large scale everywhere and have even become part of regular practice." Admitting that this was not the case "at interior posts where conditions are not always favorable," Kok nonetheless gave as the principal reason for the wide use of transfusion "the efficacy of the procedure, the simplicity of instrumentation, and the lack of specialized and expensive medicines during the war."[51]

The thesis provided few details about location, except for one reference to Katanga. It concentrated instead on practical techniques such as obtaining donors, in which case Kok went first to the immediate family or friends of the patient, with a preference for young females "who agree more voluntarily than the men," and if unavailable then made a request to infirmary personnel. Compatibility testing was done by the simple mixing of blood drawn from donor and patient. Wasserman tests were done if time permitted, and 250 cc of blood was typically drawn into a mixture with sodium citrate. Blood was given to the patient in the sickbed, and the author stated, "in over a hundred blood transfusion done in ten years I never encountered serious shock." He described the risk of transmitting various diseases, with some (e.g., tuberculosis, sleeping sickness) being more serious than others. By taking precautions such as examining potential donors, he concluded that "the danger of transmitting illness by blood transfusion is not very serious." As to the illnesses most frequently and effectively treated by transfusion, the first was anemia, the most common cause of which was hemorrhage during a difficult childbirth; next was toxic anemia from worms; and, behind that, anemia from advanced cases of malaria. The author's ultimate conclusion: "Blood transfusion, despite the difficulties inherent in the native setting, . . . can be used more with very satisfactory results."[52]

Early Transfusion in the British Colonies

There was far more variety in the number of British colonial holdings in sub-Saharan Africa, but like the Belgian Congo there was a similar pattern in how new health facilities were developed. This development included initial investments before the First World War that followed colonial interests at ports, capitals, and business enterprises. Medical missionaries were also active, but unlike the Congo, there were significant settlers in Kenya and southern Africa.[53] Beginning in the 1920s the expansion of hospitals and

European-trained doctors followed a policy to move health facilities out of the capitals to rural areas where missionaries had mostly been providing Western medical care. "A government hospital is a tangible sign of Government activities which is understood by every native," argued J. L. Gilks, principal medical officer for Kenya in his 1921 annual medical report.[54] "It is a fact which cannot be gainsaid, that the provision of medical attendance, even of the crudest and most primitive description, is the best form of advertisement for any form of activity among natives." In 1925 there were twenty-three colonial medical service doctors in Kenya and twenty-five in Uganda. Ann Crozier's study found that a total of 424 colonial service doctors had served in Kenya, Uganda, and Tanganyika by 1939.[55]

A very rich source of evidence about transfusions in British African colonies before 1939 comes from the British Red Cross, which created branches in the colonies. In the settler colonies of Kenya and Southern Rhodesia, for example, blood donor panels, or lists of donors, were established in the 1930s as a way of obtaining more reliable sources of blood for transfusion both for Africans and Europeans. This method of donor recruitment was developed between the wars in the large cities of Britain, France, and the United States,[56] where hospitals compiled lists of volunteers who were pretested for blood type and screened for illness. They were to be called, even on short notice, to have blood drawn when a transfusion was needed. Not only was this system inspired by the need for a more reliable source of blood for transfusions, but volunteering on these panels was seen as an activity to draw volunteers to help start Red Cross branches in the British colonies. The most obvious significance of establishing blood donor panels was to stimulate interest in transfusions by Western doctors already in place and with knowledge of the procedure. In effect, this was a case of supply stimulating demand.

The pioneer of this model of blood donor service was Percy Oliver, who was invited to give a talk at the 1930 British Empire Red Cross Conference, held in London, relating his experience of over a decade in that city. Shortly after the First World War, Oliver, his wife, and other members of a local Red Cross division in the London neighborhood of Camberwell answered a chance call from a local hospital to give blood for a transfusion. Until then, hospitals had relied on nurses, orderlies, or other hospital staff to serve as donors when no family member was available who matched the blood type of a patient in need of a transfusion. Oliver contacted other members of his neighborhood division, and over the years hospitals in London came to rely on this ready source of blood for transfusion. Oliver and his wife were called when the need arose, and the volunteer was sent to the hospital, where blood was drawn and given to the patient. Oliver reported that the organized service, which began in 1921, provided over 1,360 donations in 1929.[57]

The London conference was an opportunity for representatives from dozens of branches of the British Red Cross in colonies and dominions around the world to meet as well as to hear speeches and reports of activities. Oliver was one of the first plenary speakers, because his London Blood Transfusion Service was a very successful and highly visible program of the British Red Cross. At the 1930 conference he recommended work with blood donors "to all delegates as a very fine form of service for Red Cross members," but he warned them not to serve simply as a channel to recruit donors to be placed in the hands of the hospitals. His "bitter" experience was that the chapter needed to act "as a buffer between the institutions and the donors, to protect their interests."[58]

It did not take long for members in both Kenya and Southern Rhodesia, where Red Cross branches had been established only a few years earlier, to start blood transfusion services. The 1932 annual report for Kenya stated, "A blood transfusion service has been organized and has a panel of 24 donors, including 10 members of Toc H [a service club started by WWI veterans], for whom lectures on the subject were arranged." The same year the Southern Rhodesia branch reported, "a Blood transfusion service has been organized and a number of VAD [Voluntary Aid Detachment] members have enrolled as donors."[59] Indirect evidence of blood transfusions in Northern Rhodesia is contained in a March 1931 administrative report from the commissioner of Northern Province about *banyama,* or vampire men, the rumors about which were being fueled by appeals for blood donors and transfusion in the province.[60]

As the numbers indicate, this was a small start, and in subsequent years the numbers did not grow very quickly. The Kenya Medical Research Laboratory, in Nairobi, reported annual blood group tests in the 1930s of between ten and thirty individuals each year. The Rhodesian Red Cross branch stated in 1939 that the number of volunteers had risen to 903, with 650 of them grouped. "No life will be lost for lack of a willing donor," the 1939 annual report proudly boasted.[61] In fact, that same year Southern Rhodesia proclaimed with much fanfare the establishment of a "National Blood Transfusion Service," including a new building. This was, of course, a premature and hollow boast, partly because of the limited numbers, but also because it ignored the problem of saving the lives of all Africans. In any event, the war quickly put an end to such plans, yet this is at least an indication of the technical feasibility of a blood transfusion service in Africa.

These examples are telling of the practice of transfusion that can be found in the published literature and unpublished colonial reports before the Second World War. Yet they are not complete. For example, they do not discuss transfusion in French colonies, which will be covered in the next chapter, nor do they include unpublished or otherwise unrecorded individual cases, like Lejeune's patient whose desperate conditions also prompted transfusions to save lives. In the end it is impossible to know the full extent of transfusion during this period because it simply was not always judged worthy of reporting. In fact, regular inclusion of transfusion in French colonial medical reports did not begin until 1955. Only rarely did a hospital or colonial report mention the establishment of a transfusion service, as in 1949, when the two big hospitals of Dakar did so. There were also reports of a handful of transfusions in surgery services of hospitals in the French Congo in 1933 and 1934.[62] Another example of an exception that demonstrates the case in point, comes from Sierra Leone, a fairly small British colony (1931 census of 1,768,480), with few Europeans, and not particularly noted for significant investment in health or other Western development. Yet in 1936 the annual report of the pathology laboratory of Connaught Hospital in Freetown mentioned grouping seven blood donors (six African), a figure that rose to thirty-six (twenty-five African) in two years. Similar scattered examples show the widespread ability, even if limited in practice, to do blood transfusion in sub-Saharan Africa before the Second World War.[63]

In Uganda, for example, blood grouping was first reported by the Kampala Medical Laboratory in 1931. There were similar reports from Tanganyika in 1932 and the Gold Coast in 1935.[64] The 1939 annual health service report for French West Africa stated there were 140 blood group tests by hospital laboratories (24 for Africans), a figure that rose to 891 the following year (813 for Africans).[65] In the Belgian Congo, similar sources reported 18 blood group tests in 1929 in Katanga, rising to 46 (22 for Africans) in 1939. Similarly, the bacteriology laboratory in Léopoldville reported 75 blood group tests (27 for Africans) in 1937.[66] Figures from these reports are, therefore, undoubtedly a low estimate of transfusions done, since they could be and certainly were also done using blood donated by a relative or member of a hospital staff, without assistance of donor panels, and without being reported.

TABLE 1.1. **Transfusions reported (more than 10 annually) in African colonies by World War II**

Colony	*Date and notes*
Belgian Congo	1924 Haut Katanga, 300 patients; 2 other locations by 1940
Uganda	1931 first blood-grouping reports from Kampala
Kenya	1932 Nairobi, 24 donors
Tanganyika	1932 first blood-grouping reports
French Congo	1933 Brazzaville
Ethiopia	1935 first report of transfusion service in Addis Ababa
Gold Coast	1935 first blood-grouping reports
Sierra Leone	1936 first reports, 38 in 1938
Rhodesia	1939 report of 903 donors
Senegal	1940 report of 813 blood groupings, French West Africa
French Soudan	1941 300 blood-grouping reports

Sources: D. Spedener, "Le traitement des pneumonies des noirs par transfusion de sang des convalescents," *Bulletin médical du Katanga* 1 (1924): 234–38; Germond, "Statistiques des cas de pneumonie traités par transfusion de sang de convalescents," *Bulletin médical du Katanga* 1 (1924): 243; Uganda Protectorate, *Annual Medical and Sanitary Report,* 1931, 49; Kenya Colony and Protectorate, *Medical Research Laboratory Annual Report,* 1933; Tanganyika Territory, *Annual Medical and Sanitary Report,* 1932, 67; Inspection générale du Service de santé, AEF Colonie du Moyen-Congo, "Rapport annuel," 1933, 111, and 1934, 124, box 117, IMTSSA; R. Ghose, "History of Blood Transfusion in Ethiopia," *Ethiopian Medical Journal* 31, no. 4 (April 1963): 208; Gold Coast Colony, *Departmental Reports,* 1935–1936, 44; Sierra Leone, *Annual Report of the Medical and Sanitary Department,* 1936, 51; *Report of the British Red Cross Society for 1939,* 92, 95; AOF, Service de santé, *Rapport annuel,* 1940, 79; AOF, "Inspection générale des services sanitaires et médicaux, Rapport annuel 1941," 115, box 4, IMTSSA.

The records between the wars, therefore, show that all conditions existed in sub-Saharan Africa that were necessary for blood transfusions to take place: availability of donors, willing patients, and technical ability to do transfusions. They also suggest that the numbers were limited, primarily by the availability of Western medical doctors and facilities to do transfusions. There is also a hint of how innovation took place, usually through connections to knowledge and resources outside the established colonial medical structures (e.g., Red Cross, universities, mining). With this overview of the interwar period as a base of reference, the changes can be better appreciated that took place during and after the Second World War that dramatically spread and increased the use of blood transfusion in Africa.

2 BLOOD TRANSFUSION FROM 1945 TO INDEPENDENCE

There was sufficient Western medical infrastructure to make blood transfusions possible in Africa between the world wars, but this did not immediately lead to large numbers of transfusions. The rapid increase came after the Second World War, for a number of reasons. The explanation of how this rapid growth happened in most colonies is best understood by the changes in general conditions that increased the number of hospitals and brought more doctors to Africa who were able to use transfusions.

The period after 1945 in the history of modern health care in Africa is usually subsumed together with the rest of colonial rule and contrasted with the dramatic growth of health facilities after independence. Compared to the interwar colonial period, however, there was a sharp increase in hospital construction and training of medical personnel after 1945. Construction of new and modern hospital facilities after the Second World War was not only the most visible evidence of these investments but also the one with the greatest direct impact on transfusions. In French West Africa, for example, thanks in part to the FIDES (Fonds d'investissements pour le développement économique et social), created in 1946, the number of "general hospitals" rose from two in 1938 (both in Dakar) to twelve in 1952, with a corresponding rise in the number of hospital beds from 1,630 to 3,810. In Belgium the Van Hoof–Duren Plan of the 1940s called for the creation of a medical-surgical center with 100 to 150 beds in each of the 120 administrative sectors of the colony.[1] Similar projects were supported by the Colonial Development and Welfare Acts of 1940 and 1945 in Britain, such as a ten-year plan in 1946 for health in Nigeria that established a medical school and university hospital at the University of Ibadan in 1948.[2]

The growth of health facilities after the war created more places where transfusions took place, while at the same time changes in techniques during the war made it even easier to practice them. Among the most important innovations was the ability to store whole blood as well as to separate and freeze-dry plasma. Although the latter technique was never widely used in Africa, in those places where electricity and refrigeration came to hospitals, it was feasible to have "blood banks" (in the sense at least of being able to store blood). The latter by no means replaced the practice of drawing blood from a donor at the time of transfusion in many parts of Africa, but the overall result of changes in transfusion practice during the Second World War was to make its use in treatment of patients much more routine. This was reinforced by changes in training and practices in Europe that made doctors who came to Africa after 1945 much more familiar with transfusion.

Conditions in Africa during and after the Second World War

With one important exception, the immediate effect of the Second World War was to hinder the use of transfusion in African colonies because resources were diverted elsewhere. In addition, there was almost no fighting in sub-Saharan colonies that might have prompted the need for transfusion, and generally the region was too remote to be a source of blood for troops fighting elsewhere. Kenya reported limited blood donations for military and civilian patients during the war, but there were no programs in British Africa, such as were instituted in India or Australia, whereby large-scale blood collection services were established to support the fighting front.[3]

The exception came toward the end of the war in French West Africa. After the Allied landings in North Africa in November 1942, the French set up a transfusion service in Algiers, and in 1944 Gaston Ouary was sent there from Senegal to learn the new techniques. Ouary was a surgeon who had occasionally given transfusions before the war at the so-called Hôpital indigène (later Hôpital Aristide le Dantec) in Dakar.

More will be said later about how this visit changed blood transfusion in Senegal and the rest of French West Africa. Of note here is how much Ouary was immediately impressed by the new techniques he saw in Algiers. In the report he filed with authorities upon his return, in November 1944, Ouary compared what he had just learned in Algiers with prewar transfusions he had done. "Transfusion then," he explained, "gave the impression of a minor surgical intervention with all the necessities implied." A syringe was used to withdraw the donor's blood, which was then immediately given to the patient. The French called this procedure arm-to-arm transfusion, requiring the donor to be next to the patient. Ouary explained the limits imposed by this procedure: "Transfusion is thus a veritable minor surgical intervention, possible only in well-equipped health facilities by a competent doctor, most often by the surgeon on duty. One or both must devote a rather long time for preparation and execution, which would not be a major inconvenience if it was the only urgent task to accomplish."[4] Thanks to new techniques developed for the much larger scale of blood transfusion during the Second World War, Ouary went on to explain what these techniques permitted:

> The apparatus today permits transfusions almost as easily as an intravenous injection of artificial serum. It requires a sterilized bottle containing an anticoagulant solution of sodium citrate which is attached to sterilized tubing for the collection and injection of blood. An essential feature is that the injection tubing always includes a filter required to prevent small clots. This filter was not part of earlier apparatus. . . .
>
> In sum, the technical progress today permits numerous transfusions, easily and rapidly in any location, because it has become possible to store blood in one form or another as well as to transport and inject it without complicated equipment.[5]

Doctors and Decisions about Transfusions

Even with these technical improvements, in the end the decision to do a transfusion was like the decision to use any scarce Western medical resource in places such as mid-twentieth-century Africa. Doctors still faced "urgent tasks" with only limited resources to accomplish them, and there was no obvious answer to the question of whom or what to care for first. For example, when colonial powers decided to build expensive state-of-the-art Western hospitals after the war, they justified it by the need to set standards high if medical care in Africa was to be taken seriously by Western medicine. The common counterargument was that the money would help far more Africans if invested in more facilities that were less expensive.[6] Likewise, a doctor in a regional hospital could do hundreds of surgeries, with only basic anesthesia and antisepsis. Adding the ability to do transfusion could save lives, but so, too, could doing even more operations that did not require it.

Given these possibilities, then, perhaps the most crucial change after 1945 was that more European doctors went to Africa who were likely to be trained in the use of transfusion. This followed from a variety of underlying developments, including growth of health infrastructure in the colonies, growing demand from increasing population pressure, and the surprising postwar economic recovery of Europe, including expanding medical training. By the time of independence it is estimated that there were 450 Western-trained medical doctors in Uganda and 750 in Kenya.[7] The figures for the Belgian Congo were 731 doctors in 1959 (mostly with the government, but about one-third employed by missionaries and private companies), working in 422 hospitals (1957 report) averaging over 110 beds.[8]

The result was that even if they had no plans upon arrival to devote the time and resources to transfusion, these doctors could be persuaded to do so by something at the local level as simple as the availability of blood or the visit of a guest doctor who demonstrated new techniques. On the broader level, when colonial health authorities invested in large modern hospitals in the capitals of Africa, they were equipped with the latest facilities, including operating rooms, plus support services for radiology, anesthesia, and transfusion. Once a blood service and accompanying blood banks were established, their use quickly spread as people came from far away to take advantage of them. Even though doctors in the provinces did not set up their own service, they referred their patients to larger hospitals with the resources for transfusion until the smaller hospitals eventually made arrangements to do it themselves.

In the British and Belgian colonies, there was an outside stimulus to the introduction of blood transfusion: branches of the British and Belgian Red Cross. Because of their experience in collecting blood on the home front during the Second World War, national Red Cross societies all over the world became leaders in adapting their expertise in blood collection to peacetime operations: recruiting blood donors and in some places, processing blood for transfusion. This was the case in the United States, many European countries, Canada, and Australia, to mention just a few examples.[9] The Red Cross expertise was transferable to the colonies, where even though transfusion remained a hospital operation, Red Cross volunteers certainly made it easier to begin or expand transfusion by helping assure adequate donors and in many cases providing funds for equipment and supplies to store blood. This was less the case in the French colonies, because in France a national transfusion service emerged after 1945 out of collaborative efforts between the hospitals and governments dating back to the interwar years, with little or no participation by the French Red Cross.[10]

All colonial medical department directors were overwhelmed by the health problems in their districts. Moreover, their budgets were small, and requests for additional funds exceeded the resources and competed against one another to make services available to meet basic medical needs. As a result, viewed from the colonies, an organization like the British Red Cross held out the promise of a significant source of volunteer staff time to recruit donors, not to mention funds for such things as transportation, equipment to draw blood, and refrigerators to store it for transfusions. The Red Cross also enjoyed a formidable reputation for beneficence that bolstered confidence in any new scheme. Thus, in whole colonies such as Uganda, Northern Rhodesia, and the Belgian Congo, the Red Cross was asked to run the transfusion services, at least initially.

Despite these immediate advantages, there was a condition imposed by Red Cross involvement in blood transfusion that prompted an ongoing debate and controversy: insistence that blood donation be voluntary, that is, with no remuneration for the donor. This had become part of the ethos of the transfusion service in Britain from its start, after the First World War, and was especially championed by its founder, Percy Oliver (see chapter 1). It spread to other European Red Cross societies involved with blood collection, as in Belgium, the Netherlands, and Switzerland, where they eventually ran their countries' blood programs.[11]

This ethos did not, however, take root automatically in African societies. As will be seen, it was difficult to find adequate numbers of Africans to give blood on an anonymous, voluntary basis. When demand grew for the procedure after the 1950s and 1960s, hospital transfusion services had to adopt other means of securing blood. This was done either by direct remuneration, or by requiring patients to find a family member or friend either to be the donor for the patient or give blood as replacement to the blood bank. In addition, almost all donors were given refreshments, cigarettes, and sometimes cash. As a result, transfusion in most African countries was hospital-based by the 1970s, except in such places as Senegal and Uganda, where the newly independent countries continued and expanded the centralized blood services created during the colonial period. This meant that each hospital found its own source of donors to give blood on call or to donate regularly to a blood bank if the hospital had storage facilities. Only later and with outside financial assistance, usually prompted by a crisis or disaster, were independent African countries able to implement the centralized model of blood supply using anonymous, voluntary donors.

The Organization of African Blood Transfusion Services: General Trends and Periods

Before 1945 blood transfusion was organized in Africa by hospitals. It was decentralized, and transfusions depended primarily on the available facilities and the doctor's knowledge. This favored transfusion at bigger hospitals in capital cities where there might be three or four doctors and at least one surgeon, or hospitals with special outside links, such as ones supported by the University of

Louvain and the Union minière in the Belgian Congo. Likewise, the practice of transfusion might be started in a hospital because a doctor who had practiced transfusion at one location might bring that experience and repeat it at a new hospital assignment. This was the case with Joseph Lambillon when he moved from Kivu to Léopoldville in the Belgian Congo health service during the Second World War, and also with Gaston Ouary when he moved from Dakar to Brazzaville in the French colonial health service after the war.[12] That did not necessarily guarantee the overall increase of transfusion, since after a practitioner moved, his successor might not be knowledgeable or interested in continuing to do transfusions. Thus, when Lambillon left the Kivu hospital at the end of the Second World War, his transfusion instruments lay idle until the early 1950s, when a new doctor, Louis Legrand, arrived from Brussels who was schooled in newer transfusion techniques that he introduced.[13]

In addition to the doctor's decision to use transfusions, the selection of donors in this initial period also influenced whether the procedure was done in a particular setting. For example, in 1940 Lambillon stressed the possibilities of blood donation from recovering patients in African hospitals, but more typically family members were asked to donate. As to the uses of transfusion, there was some experimentation with transfusion for pneumonia as early as the 1920s in Katanga, because of the high incidence of that disease among mine workers, but more typical were surgery cases and difficult obstetrical deliveries. The experiment with anemic infants at Kisantu Hospital in the Congo in the early 1940s proved to be the precursor of a practice that became more widespread and particular to the African setting in the 1950s and 1960s.[14]

To summarize, by 1939 transfusion was known to doctors in most capitals and big hospitals in sub-Saharan Africa. Connections back in Europe and the small world of colonial medicine facilitated this. The extent to which transfusions were done varied depending on local circumstances such as the interest of doctors and surgeons or the existence of a Red Cross branch.

The policy decisions and other developments that led to widespread introduction of transfusion after the Second World War also brought an attempt to centralize transfusion services. Thus, when a new hospital was built in the 1950s, as in Ibadan, Nigeria; Kampala, Uganda; and Lomé, Togo, or an existing one was enlarged, especially with a surgery wing, as in Nairobi, it typically included the standard services for modern operations, such as expanded laboratory facilities and a blood bank.[15] Because this gave big hospitals, usually in the capital, the facilities that other hospitals did not have, their blood collection, testing, and banking facilities often became at least citywide services and, where feasible, sometimes reached nearby district hospitals. In large and relatively prosperous colonies such as Kenya and Uganda, the transfusion services and laboratories served other hospitals as far as transportation of blood would allow. The Dakar federal transfusion center went to the furthest extreme when it attempted to provide blood not just for Senegal but all of French West Africa.

Following the Red Cross model in British colonies, blood was usually expected to be donated voluntarily during this period from the Second World War to independence, but French colonies generally followed the metropole model, where the government set a price to compensate for the effort to make a blood donation.[16] There were pressures, however, that produced a mixture of paid and voluntary donation everywhere. In some of the British colonies, for example, there were hospitals that did not rely entirely on Red Cross voluntary donors; thus there was already a mixed approach before independence.[17] Likewise, both the Red Cross volunteer system of collection and the Dakar center recruited unpaid donors from the Westernized African classes and workforces: army personnel, civil servants, and factory workers, but above all older schoolchildren and prisoners. The practice in Senegal was that if donors came to give blood at the transfusion center, they were paid for their trouble and fed, but donors at mobile units on-site were not. The question of who donated and who used blood will be examined in greater detail in chapters 4 and 5. Both groups grew significantly during this time period. The most important categories of patients receiving transfusions were general medicine (including accidents and emergencies) and surgery, along with maternity services and pediatrics if these specialties were available.

After independence, the organization of transfusion services entered a new phase, with most countries accelerating expansion by building provincial hospitals to serve regional needs better. This was also in response to the higher cost and slowness of transport that occurred in the centralized model. Other countries, which had never been able to centralize, such as Nigeria and Congo/Zaire, left it to the local hospitals to arrange for their own transfusion services, sometimes with the assistance of the Red Cross, sometimes with a paid service, and sometimes both. Thus, after independence there was a general swing away from centralization and its high costs, toward a middle position of mixed organization with limited regional services at best, and hospital-based means to supplement or complement blood collection and testing. In general, this move was driven by the continued increase in the use of transfusion and the corresponding need for more donors, which had accelerated in the last ten to twenty years of colonial rule.

Transfusions in French African Colonies after the Second World War

In most African colonies, the Second World War diverted resources elsewhere and reduced the practice of blood transfusion. The major exception to this, as mentioned above, was Senegal, where shortly after the Allied invasion of North Africa planning began for the Pasteur Institute in Dakar to collect and ship blood to the front. This development had significant repercussions for the organization of transfusion services, not just in Senegal but all of French West Africa.

In September 1943 the Dakar Pasteur Institute was instructed to prepare test sera for blood group determination of European and African troops stationed in Dakar and the Senegal-Mauretania colonies. In addition, the Pasteur Institute was to ship test sera to all colonies in French West Africa. By the end of 1943 over six thousand vials were prepared and 449 Africans and 166 Europeans had been tested.[18] In 1944, Gaston Ouary, a surgeon in the colonial medical service at the African hospital in Dakar, was sent for training to the blood transfusion center established in Algiers by Edmond Benhamou.[19] After Ouary's return, he and two other colonial medical officers, Yann Goez and Jacques Linhard, secured the equipment necessary to set up a service at Dakar, including writing a manual for training personnel to draw blood and perform transfusions. In February 1945 these personnel, under the direction of the Pasteur Institute, began their transfusion service in an American army barracks on the outskirts of Dakar. By the end of the war enough blood was collected to provide over 225 liters for transfusion, mostly in the form of plasma but also some whole blood that was shipped to troops in Italy.

This wartime development also had an immediate impact on civilian blood transfusion, because once a source of blood was available, it was also used by the main civilian hospital in Dakar and the military Hôpital principal.[20] In fact, French colonial authorities were quite aware of these extended benefits from the start. Thus, when Ouary returned from his training in Algiers, the transfusion manual he wrote with Goez and Linhard in November 1944 was not just for wartime use.[21] As the head of the French colonial health service, Marcel Vaucel, stated in his introduction, the authors of the book had a double purpose: "to describe the new technique for their distant comrades, [and] to expand the uses of blood transfusion in tropical locales." Benhamou repeated this in the conclusion to his preface: "We are sure that blood transfusion in all its forms (fresh whole noncitrated blood, stored whole blood, blood products) has a large future in our colonies, and that the notes so brilliantly edited and perfectly illustrated by Médecins-Commandants Ouary and Goez, and Médecin-Capitaine Linhard, will significantly aid in the diffusion and expansion of this heroic therapy, in war as in peacetime."[22] And, as the authors themselves put it, "We thought it useful to make these [scientific and technical developments] known to our comrades in the empire, who work without access to publications and who find it impossible to follow the medical progress achieved during this war."[23]

The report that Ouary wrote to his superiors upon his return indicated the implications for infrastructure that predicted some of the subsequent developments of transfusion services in most African colonies and independent countries. "Today those who must care for the wounded demand larger and larger quantities of blood. Such an increase in transfusion has necessitated the creation of a new organization." He pointed out that the Algiers center included:

- a laboratory to prepare the different types of blood, furnished by its own collection sources, mobile teams both lightly and fully equipped, and secondary fixed centers
- a warehouse to provide equipment and biochemical supplies
- a training center for reanimation-transfusion teams[24]

Following the plan of the Colonial Health Department, the Pasteur Institute continued the blood collection service in Dakar after the war; and although the amount of blood collected dropped to less than thirty liters in 1946, donations steadily grew thereafter.[25] In 1949 the two large Dakar hospitals (Hôpital le Dantec and Hôpital principal) had organized transfusion services, the one in le Dantec being housed in a new surgery wing completed that year, with twenty-four beds dedicated to "reanimation." Louise Navaranne, a doctor who accompanied her husband, Paul, to Dakar when he was assigned to the surgery service of le Dantec, directed the reanimation center. In 1950 she reported almost four hundred transfusions with whole blood and plasma supplied by the transfusion service.[26]

At this same time, credits were voted to establish a federal transfusion service that opened in 1951 to serve all colonies in French West Africa.[27] In a letter to the governor general of French West Africa, the director of public health for the federation, Léon Le Rouzic, gave four reasons for the creation of the federal transfusion center, some of which proved to have clear foresight, combined with others that never saw the light of day. They were:

1. The important increase in the number of serious accidents occurring each day in Dakar and its environs.

2. The capital of AOF [l'Afrique Occidentale Française] has an airport that has become a crossroads of international airlines, and health facilities must possess a maximum of resources in case of an airline accident. It is noticeable that foreigners are concerned about the means at our disposal in this regard.
3. This facility will become part of the health facilities of greater Dakar.
4. The transfusion center will be a federal facility, with blood and plasma capable of being sent at any time to facilities in the interior by regular airlines or planes (military or civil) required for this.[28]

Linhard, who had trained in obstetrics at Bordeaux a few years following Ouary in the 1930s and was coauthor with him of the transfusion manual for use in the colonies, became the first director of the transfusion service in Dakar. The reports of the service quickly showed that sufficient donors were found that met the greatly increased demands for transfusion. After six months of operation, the Dakar center reported 3,508 donations of 300 to 350 cc each from almost two thousand donors. Of these 1,384 were Africans, mostly civil servants (the Europeans were military), and none were women. Within three years, according to the center's 1954 report, it drew blood from over twelve thousand donors, all but five hundred of whom were Africans.[29] Significantly, and uniquely for sub-Saharan Africa, the Dakar transfusion service had facilities for freeze-drying plasma, which was useful for shipments to the other colonies.[30] The increased supply of blood and plasma made it possible for le Dantec Hospital to increase its blood transfusions dramatically, according to the use of blood products at the hospital before and after the opening of the new transfusion center.[31]

Most of the blood and plasma were distributed by train to Senegal and the French Sudan, and by air to the major cities of French colonies throughout West Africa: Bamako, Conakry, Abidjan, Niamey, Ouagadougou, Lomé, and Cotonou, as well as Douala, Cameroun (a UN protectorate administered by France). This regional approach, and the possibilities it implied for centralized services and quality control, proved to be exceptional and temporary. For reasons

TABLE 2.1. **Blood and plasma use, le Dantec Hospital, Dakar, 1950–52**

Year	*Whole blood (250 cc units)*	*Plasma (350 cc units)*
1950	180	207
1951	384	286
1952	1,146	675

Source: "Rapport annuel, Hôpital Central Africain," 1950–52, box 32, IMTSSA.

of cost, increased demand, and the growing political-independence movements, separate transfusion services were soon created in each colony.

Before the establishment of the blood collection service in Dakar, there were reports of transfusions in the French Congo as early as 1933 and 1934. Additional evidence shows transfusions there in the late 1940s as well.[32] Likewise, there is indirect evidence of transfusions from lab reports of blood group testing in French West Africa in 1939, most likely from the Dakar hospitals but also in the French Sudan between 1940 and 1945. Gabon reported blood group tests in 1950 and 1951, and, along with Togo and the French Congo, it established agreements in 1950 on the price paid for blood given by local donors, in accordance with the national agreement negotiated in 1949 between the Ministry of Health and the Fédération nationale des donneurs de sang de France et d'Outre-mer.[33]

These accounts suggest a pattern outside Dakar that followed the interwar record in Belgian and British colonies, where the individual interest of doctors or other circumstances determined the use of transfusions. And colonial health services moved doctors around fairly regularly. Thus, for example, the transfusions done in the French Congo in 1950 followed the appointment of Ouary as surgeon at Brazzaville Hôpital général after he left Dakar. In 1955 he was chief of the surgery service in Tananarive, Madagascar.[34] Likewise, when Togo completed construction of a new hospital in Lomé in 1954, Amen Lawson headed the bacteriology department, and unilaterally started a paid blood donor service, because it was much cheaper and more responsive to immediate needs than service from Dakar.[35]

TABLE 2.2. **Total blood units supplied, Centre fédéral de transfusion (Dakar), 1950–58**

Year	*Total donations*	*African donors*	*Whole blood (250cc)*	*Plasma (350cc)*	*Dried plasma (liters)*	*Total units*
1950	—	—	530	389	136	919
1951	6,910	6,070	2,854	3,089	1,081	5,943
1952	11,273	10,393	2,523	5,280	1,848	7,803
1953	9,588	9,158	2,366	3,680	1,288	6,046
1954	12,245	11,797	3,414	5,394	1,888	8,808
1955	11,292	10,842	6,342	6,829	2,390	13,171
1956	11,809	11,164	9,091	5,509	1,928	14,600
1957	14,018	12,795	12,484	5,260	1,841	17,744
1958	14,181	11,451	12,973	4,314	1,510	17,287

Note: Units are either 250 cc whole blood or 350 cc plasma. Three blood products were produced at the center: Whole blood, liquid plasma, and dried plasma.

Source: Unclassified records, CNTS Dakar.

The reports of overall blood and plasma production through 1958 give an indication of the number of transfusions in French West African colonies for which the blood was supplied.[36] Of note was the rapid growth but quick leveling off of donations and units produced, likely due to costs, plus the very high rate of blood donation by the local population. African donors made up the vast majority from the start, and there was a steady growth of whole-blood collection. Plasma remained a significant portion of production but leveled off after 1954.

Detailed reports have not been found about where shipments went from the federal transfusion service in Dakar, and it was only in 1955 that an official category was created for blood transfusion in the annual French colonial medical reports. Nonetheless, based on intermittent reports, it is clear that transfusion was widely used, and in some places regularly established, in French West and French Equatorial African colonies by 1956.[37]

Given the widespread ability to do transfusions, it follows that the main initial effect of the Dakar service, as far as West African and Cameroon colonies are concerned, was to expand the practice. In other words, this was likely an unusual case of increased supply stimulating demand. Then by the late 1950s, as demand in the colonies began to exceed the ability of Dakar to supply blood and blood products, especially at a reasonable price, the hospitals in the other

TABLE 2.3. **Transfusions reported with whole blood and plasma, French African colonies, 1955–56**

Colony	1955	1956
French West Africa		
Ivory Coast	79	695
Dahomey	—	286
Guinea	164	—
Upper Volta	25	40
Niger	148	182
Senegal	—	—
Dakar hospitals	577	867
Other	1,054	293
Togo	—	180
French Equatorial Africa		
Ubangi-Shari	37	33
Congo	55	116
Gabon	63	22
Cameroon	—	273

Source: Annual medical reports for each colony, 1955, 1956, IMTSSA.

colonies developed their own local sources. Sometimes this was done publicly and openly, as in 1957, when the Ivory Coast officially voted to create its own blood transfusion service. The minutes of the territorial assembly reported the health minister's testimony: "A blood bank is indispensable to the colony at this time because of the increase in patients who can benefit from whole blood and whose needs are always urgent in nature. The center in Dakar, he said, has prices that are too high. He has been forced on several occasions to order blood directly from France."[38] Other colonies, such as Togo, did not require such dramatic action. A hospital might simply ask a patient's family to find a donor, or develop a more systematic way of insuring blood for transfusion quickly and affordably. In any case, the trend was clearly toward a decentralization of blood collection that foreshadowed the pattern for the period of independence.

The evidence about blood transfusions in former French African colonies is very broad but unfortunately also very shallow. It provides a fairly complete account of when and where and how many transfusions were done over a large part of West and Central Africa, but there is less evidence about who gave blood and for what purposes. One particularly intriguing feature is that the population of the Dakar region was essentially donating blood for all of West Africa during most of the 1950s. Moreover, the blood donation rate (for example, 14,181 donations in 1957 for a population of 234,500) was 6,047 per hundred thousand, easily the highest found anywhere in Africa and well in excess of the two-to-four-thousand per hundred thousand rate that became the standard for donations in Europe and North America. The reason why so much blood was donated is only partly explained by tradition and military troops stationed in Dakar. It was likely also the result of the French policy of a 500-franc (CFA) payment for a donor's time, plus refreshments (a sandwich and a drink).[39]

More will be said later about who donated, but overall these French records emphasize the colonial administrative part of the story. Of note here is that with only a few exceptions, the doctors and administrators remained largely anonymous, thanks to the centralized bureaucratic system of French reporting. Fortunately there is much richer evidence about the history of blood transfusion in the British and Belgian colonies, because of the participation of the Red Cross societies. Although their records are also biased toward documenting the work of Europeans involved, their detail permits a better indication of the Africans who were the patients, donors, and part of those who organized and administered the transfusions.

Transfusions in the British African Colonies after the Second World War

Blood transfusion in the British African colonies, as in the French colonies, ultimately depended on a doctor's decision to use the procedure for patients. That decision, however, was strongly influenced by the state of local health facilities, including the existence of a hospital and a readily available blood supply. After the 1920s any doctor in Africa who was intent on doing so could give a blood transfusion to a patient in a Western hospital by finding a donor from the hospital staff, the patient's family, or like Lambillon, even from convalescing patients. This search could require some effort, and it stands to reason that if a supply of blood were available, doctors would be more inclined to give transfusions to patients. In this scenario the doctor would be the bottleneck limiting blood transfusions. If patients were in need of transfusion, a service organized for recruiting donors and processing blood might persuade a reluctant doctor to give transfusions. In British African colonies, local Red Cross branches frequently served that purpose in facilitating transfusions.

When the government of French West Africa, for example, took steps to continue a blood supply after the Second World War, it stimulated the use of transfusions in hospitals, not just in Dakar and Senegal, but elsewhere in West Africa. This approach, however, differed significantly from what happened in the British African colonies. For, unlike the French colonies where the government took the lead, it was the local Red Cross branches who took the initiative in the British colonies, either to respond to a request by a hospital or medical service to find blood donors or to initiate the idea by approaching the medical authorities with an offer to find volunteers to donate blood. As Percy Oliver pointed out at the British Empire Red Cross Conference in 1930, blood donation was "a very fine form of service for Red Cross members."[40]

Records show that Europeans in Africa began the process, and in some settler colonies such as Kenya and Southern Rhodesia, all parties (patients and donors, as well as doctors) operated in a segregated system, at least for a while.[41] But that did not remain the case for long. In places like Tanganyika and Uganda, let alone Nigeria or the Gold Coast, there simply were not enough Europeans to operate a separate transfusion system. And even in the settler colonies the increased government expenditures on health services after the Second World War meant that transfusions and other medical treatments had to be extended to African patients. The ethos, not to mention practical politics at the time, would not allow such blatant racism. Once doctors decided that transfusion was appropriate for African patients, there were not enough Europeans to serve as donors to meet the rapid rise in demand as Africans agreed to take advantage of the treatment. European patients might insist on a European blood donor, but this quickly became a marginal part of the blood transfusion service compared to the large African population in the colonies that needed and donated blood for transfusion.

In British colonies there was a scattered record of blood group testing before 1945, according to annual laboratory reports that indicate widespread but probably infrequent use of transfusion. The lack of published articles by British authors between the wars suggests no sustained attempts at treatment or service like there were in the Belgian Congo. A Red Cross chapter could organize a panel of donors on a small scale for occasional use by a local hospital, which was the case in Kenya as early as the 1930s. The Southern Rhodesia Red Cross had larger ambitions when it launched the National Blood Transfusion Service in 1939, but the plan was cut short by the outbreak of the Second World War.[42]

The introduction of transfusions on a regular basis in most British African colonies came after the war. Table 2.4 summarizes the record, drawn often from government and Red Cross reports.[43]

TABLE 2.4. **Blood transfusions reported in British colonies, 1947–62**

Year	*Kenya*	*Southern Rhodesia*	*Uganda*	*Tanganyika*	*Northern Rhodesia*	*Nigeria*	*Gold Coast*
1947	248	246	—	69	82	—	—
1948	241	—	—	153	84	—	—
1949	323	—	720	140	96	—	—
1950	421	382	720	—	140	—	—
1951	500	563	648	—	592	—	—
1952	—	1,376	—	—	150	—	—
1953	550	—	558	268	279	1,168	41
1954	—	—	556	162	—	250	335
1955	840	—	664	288	—	540	312
1956	—	—	1,269	—	—	1,603	1,323
1957	1,884	—	1,409	652	—	6,057	2,336
1958	—	—	2,726	664	—	7,361	—
1959	5,146	—	3,874	1,031	1,085	—	—
1960	—	—	5,500	896	2,219	—	—
1961	—	—	8,533	1,109	2,140	—	—
1962	11,282	—	9,000	—	—	—	—

Sources: Published annual reports of colonial government medical departments, and Red Cross reports, 1947–62, BRC London.

The early and significant use of transfusion in Uganda is clearly shown by table 2.4, as is the late start in the West African colonies of Nigeria and the Gold Coast. It does not follow, however, that lack of data means that transfusions stopped, especially after the first reports. It was much more likely that reports were simply not filed. Table 2.4 shows that by 1953 all these British colonies except one reported transfusions whose numbers were at least in the hundreds and grew at an accelerated rate during the 1950s to ten thousand or more annually in a few colonies by 1960.

The increase in transfusions in settler colonies of East and southern Africa after 1945 occurred because black Africans were included, and the Red Cross branches in these colonies were very much involved in the process. In Southern Rhodesia the Red Cross branch at first attempted to supply all blood needs, including the needs of African hospitals, by using white donors, according to the head of the British Red Cross overseas branches who visited there in 1948. By 1950, however, a separate African blood bank was established in Bulawayo.[44] The Kenya Red Cross branch reported the establishment of a blood transfusion service in 1947 at King George VI Hospital, the main hospital in Nairobi (now called Kenyatta National Hospital), with 248 blood donations reported that year. In the second half of 1948 the pathology laboratory of the Nairobi European Hospital (now the Nairobi Hospital) reported that

TABLE 2.5. **Beginning dates of Red Cross Branch Blood Collection Service in British African colonies after World War II**

Colony	Year
Kenya	1947
Nyasaland	1948
Tanganyika	1948
Uganda	1948
Northern Rhodesia	1949
Basutoland	1951
Gold Coast	1952
Nigeria	1953
Sierra Leone	1956
Gambia	1959

Sources: Joan Whittington, "Report on the Nyasaland Local Branch," June 9, 1948, Acc 0287/46 Nyasaland; Whittington, "Report on Visit to Tanganyika Territory," June 9, 1948, Acc 0287/60 Tanganyika; "Report for the Year 1948 from the Uganda Central Council Branch," Acc 0287/63 Uganda; "Report to British Red Cross from Lusaka," October 25, 1949, Acc 0076/38(1); "Miss Borley's Report," November 1950, Acc 0076/6(1); Gold Coast, "Summary Report for 1952," Acc 0287/33 Gold Coast; Nigerian Central Council, "Annual Report," 1952, 5, Acc 0076/36(1); Sierra Leone Branch Red Cross Society, "Annual Report," 1956, Acc 0076/48(2); M. D. N'Jie, "Red Cross Week, 9th–14th March, 1959," March 20, 1959, Acc 0076/21(2) Gambia, all in BRC London archives.

191 Africans were typed as blood donors for family and friends.[45] In Northern Rhodesia the Red Cross branch was asked to establish a blood transfusion service in the colony, beginning in 1950 at the African hospital at Lusaka, while in West Africa, the Red Cross involvement came a few years later and depended on relations between the local branches and hospitals.[46] To summarize, by 1953 all major British colonies in Africa had organized blood transfusion services.[47]

It is impossible in this study to provide a detailed history of blood transfusion in each colony, but closer examination of the records in Uganda and the Belgian Congo provides examples of the complexities not revealed in the broader survey of developments.

Transfusions in British African Colonies: The Case of Uganda

One of the most interesting and successful efforts at establishing a transfusion service was in Uganda, where local health authorities asked the Red Cross branch to establish a blood transfusion service in 1948. The response surprised everyone, as Uganda developed the first colonywide blood transfusion service in a sub-Saharan African state. Because of the extent of activity, Uganda also provided an early indication of who donated blood, even if the record of who received transfusions is still not very well documented.[48] The details are worth examining, not because they were typical but because they illustrate the possibilities.

On May 4, 1948, a meeting of the self-styled Sub-committee of the British Medical Association in Uganda called for "setting up a Blood Transfusion service to meet the demands for blood transfusion for all races in the vicinity of Kampala and in exceptional circumstances, in any part of the Protectorate."[49] Several features of the three-page report were telling. First, there were only four members present, although they represented the surgical staff and pathology laboratory of Mulago Hospital, the biggest and most important government hospital in the protectorate. Included was Ian MacAdam, a recently arrived surgeon who remained in Uganda until 1972 and helped build the hospital's reputation with doctors that he attracted, including subsequent Nobel laureate Denis Burkitt. Second, it is clear from the document that transfusions were already being practiced both at Mulago Hospital and Mengo Hospital, the first Western hospital established in 1897 by missionary Albert Cook in Kampala. Part of the justification for setting up the service, as the report pointed out, was that the necessary staff were already at Mulago Hospital for such things as lab work and sterilizing equipment. "For several years," reported the president of the Uganda branch of the Red Cross in 1949, Mengo Hospital "had obtained blood, in cases of dire necessity, from dressers [doctors' assistants] and students."[50]

To establish a more reliable blood supply for more frequent transfusions, the subcommittee pointed out, what was needed—in addition to refrigeration, transport, and identification of adequate donors—was an "organizing secretary." This was an overt appeal to the British Red Cross in London to supply the equipment and personnel. The May 4 request spelled out quite clearly what the duties of the secretary should be:

(a) the propaganda for Blood Transfusion
(b) accurate records of donors
(c) the organizing of blood collection
(d) the responsibility for equipment
(e) liaison between the various major hospitals

"She" the report went on, specifying the secretary's gender, "should be responsible to a joint committee appointed by the BMA."[51]

The Red Cross branch in Kampala was so enthusiastic in its desire to cooperate that it hired the wife of a physiologist at Mulago to be a part-time secretary and installed her in an office with telephone, stationery, and index cards, even though it did not yet have the resources for the other equipment. For that, the Uganda branch sent a request to London for funds, specifically to Joan Whittington, the director of the British Red Cross overseas branches, who had visited Uganda earlier in the year. While awaiting approval, the Uganda branch began publicity and lectures to educate and motivate potential donors.[52] London quickly approved the request for £750 to purchase a refrigerator, a van, donor sets, needle-sharpening equipment, bottles, and other supplies and equipment.

Some features of the new blood transfusion service, such as recruiting donors at schools, colleges, and missions, set a pattern that was followed not just in Uganda but most African colonies and countries. Other early practices, such as typing all potential donors but contacting only those with blood type O, were abandoned as soon as demand required more donors. In the first year of activity, 456 potential donors were recruited at ten schools, colleges, and the police training academy, with 228 found to be "universal donors." The initial goal was to have a supply that permitted transfusion of ten pints per week, although first reports were that the average was only six per week by January 1949. In addition to stepping up recruitment of donors, the service adopted a policy for stored blood that also gave an indication of use. The policy stated that stored blood would be used "for emergency cases only, until the day the bleeding team replenishes the [blood] bank with a fresh supply, after which the previous week's supply is made available for non-emergency transfusions such as anemia cases." To dramatize the importance of storage, the president of the Uganda Red Cross branch reported that fifty of the first eighty-one transfusions "have undoubtedly proved to be life-saving in accident cases and where cases have been suffering from post-operative shock."[53]

The demand for the transfusion service grew quickly at Mengo and Mulago Hospitals. The 1949–50 annual report for the Red Cross branch in Uganda quoted one surgeon as saying, "The field of surgery has been greatly widened by the Blood Transfusion Service and surgeons have been able to perform operations which were hitherto too dangerous or else entirely impossible."[54] The target of ten transfusions per week was met and doubled in 1949 but remained steady at around seven hundred transfusions per year until the mid-1950s, when the annual total surpassed one thousand per year. The expansion was possible because of more persuasive and extensive recruiting, including the making of a film that was dubbed into the local language and later exported to other African colonies and countries (see chapter 5).

The hospitals, meanwhile, intensified their efforts to persuade friends and relatives of patients to donate blood, not just before but after transfusion. As a 1956 Uganda Blood Transfusion Service report described it, "The Red Cross Blood Transfusion worker at Mulago Hospital, Mr. Emmanuel Muwonge, goes round the wards regularly and speaks to relatives and friends of patients needing blood transfusions. He explains the need and the technique to them, and, if they are willing to give blood he makes the necessary arrangements and assists the doctor."[55] The result, it went on, was that one patient who received three pints of blood, thus saving his life, had relatives and friends who donated a total of nine pints to the service. For the whole year of 1956, Mulago Hospital received 339 pints of blood in this manner out of a total of 1,407 pints collected for the entire protectorate.[56] That same 1956 report announced the intent of Jinja Hospital to organize another transfusion center for Busoga Province, in the east, and an Asian subcommittee was created to establish an Asian blood bank in Kampala.

The generally good records of the Uganda Red Cross document the expansion of transfusion before independence that reflected even further growth of services at provincial hospitals. By 1958 there were transfusion services at Mbale and Gulu, as well as Kampala and Jinja. In 1962 the last report of the transfusion service before independence indicated that nine thousand pints of blood were collected, still mostly in and around Kampala. By this time the majority of blood processed at the Nakasero Hill transfusion center was drawn by

TABLE 2.6. **Blood donations reported, Uganda, selected years, 1957–65 (pints)**

Region	*1957*	*1958*	*1959*	*1962*	*1963*	*1965*
Kampala	1,183	2,057	2,530	5,900		
Busoga	226	505	744	1,176	2,376	2,966
Mbale		62	105	1,077	1,702	1,914
Fort Portal			140		192	
Tororo			109	336	429	
Lira			91	188	319	
Arua			25	106	112	
Gulu		65	40	153	105	227
Ngora			30			
Soroti		17	41		586	
Masaka		20		374	570	986
Moyo				14	10	
Mbarara				62		
Moroto				22	11	
Toro				390		
Hoima					30	
Kabale					51	
					*14,000	

**Total for 1963 from separate source.*

mobile units at numerous locations including colleges, high schools, training centers, prisons, a convent, and the airport at Entebbe. The list of provincial collections was equally impressive, based on statistics for selected years between 1957 and 1965.[57]

In 1948 when the Uganda Red Cross branch responded to the call for a transfusion service, its president observed that her organization's reputation "carries considerable weight" that might help the project succeed. "It was felt therefore that if this blood transfusion service were started under the auspices of the Red Cross, it would win the confidence of the African far more than a Government-sponsored project."[58] Whether true or not, already by 1952 the newly appointed governor of Uganda, Sir Andrew Cohen, attended the annual meeting of the Red Cross branch and made special note of the success of the blood transfusion service. His speech paid tribute and also made the following prophetic observation:

> If the Red Cross and other voluntary bodies can start new services, this may become so popular and may come to be regarded as so necessary

> and essential that eventually they will become part of the fabric of the Government, and Government, whether willingly or unwillingly, will be forced to take over those services and run them themselves. That is how public affairs go. We have an excellent example of this, if I am not wrong, in the Blood Transfusion service, which is a fine service which you are now running and which eventually no doubt will become the responsibility either of local government or of Central Government.[59]

As the use of transfusion expanded both in Kampala and Jinja, as well as at up-country hospitals in other parts of the protectorate, the government (both of the colony and subsequently of the independent country) took a more enlightened approach, giving the Uganda Red Cross Society a subsidy to continue its part in the transfusion service, rather than taking it over completely as a government service. By the time of independence, however, tensions were growing between the two parties because of increasing costs and the inability of the Red Cross or government to meet them. Sue Maltby, a British Red Cross worker in Uganda, stated in a report at the end of 1959 that the blood transfusion service, "continues to expand at an alarming rate," with the result that it was always short of money. When the Red Cross asked for an increase in government subvention, it was refused. Members responded with letters written "to the minister from Lady Crawford [wife of the governor of Uganda]," followed by meetings, revised estimates, and more meetings. Only after all that, Maltby reported, was the Red Cross promised an extra £500 for 1959. And for the next year they agreed to an increase from £750 to £2,500 (current value of $60,000), "but not before we held the biggest pistol possible to their heads," she concluded.[60]

When the women in the Red Cross were the wives of the doctors using the transfusion services, these matters could be worked out "within the family." With the increasing use of transfusion and more turnover in Red Cross volunteers, however, the delicate balance between those using the blood that was provided by those doing recruiting, bleeding, testing, and storing was upset and disagreements resulted. Although the Red Cross had been offered government resources to help with the costs of its responsibilities, the Red Cross complained that the funding did not cover their rapidly rising costs. This was similar to what eventually happened in Zambia (former Northern Rhodesia) after independence.[61]

In fact, the Red Cross and the colonial government in Kenya reached a similar impasse well before independence. A transfusion service was organized in Kenya at about the same time as in Uganda and for the same reasons, but the subsequent history in this neighboring colony shows some fundamental differences, albeit with noteworthy similarities worth examining. For example, difficulties in defining responsibilities, led to an early withdrawal of the Red Cross from most transfusion activities in Kenya. This was after early and impressive interest and response in this larger and more complicated colony, thanks to the significant settler population. As in Uganda, hospitals outside Kenya's capital added significantly to the transfusion activity. In Mombasa, for example, the Red Cross division established a service for all races by 1952. Blood transfusion services were also reported in Rift Valley Province (Nakuru and Eldoret in 1953), and Nyanza Province (Kisumu in 1955). The Kenya Medical Department's annual report for 1957 indicated 125 to 150 pints of blood used for transfusion at King George VI Hospital (later Kenyatta National Hospital) monthly (almost 1,900 per year), and in 1959 the blood bank serving all Nairobi hospitals handled 5,196 pints of blood.[62] The service in Nairobi used by far the greatest resources of time and money of the Kenya branch of the Red Cross. By 1958 it ran three blood banks: one for Africans and one for Asians at King George VI Hospital, and one for Europeans at the Nairobi European Hospital. The government also increased its commitment, mainly at the laboratory of the King George VI Hospital where blood was grouped and stored.

The Kenya branch of the British Red Cross found this arrangement difficult to continue either in finding enough volunteer donors, or supplying its share of funds to match the government's contribution toward the costs of operations. Since its basic principles precluded paying for blood donation, the Red Cross found it difficult to ask patients or hospitals to pay for blood to be used for transfusions even though that might provide a source of income. Hence there was little other choice than for the government to absorb the increasing cost of the service. In November 1959 the Red Cross faced the issue when it indicated it would cut back its services at the beginning of 1960, and a new Nairobi blood transfusion coordinating committee was appointed with the medical superintendent of King George VI Hospital as its head. In an April 1960 circular to all Red Cross divisions in Kenya, the committee indicated that the Red Cross would continue to be responsible for compiling and maintaining blood donor panels of all races, including publicity to increase the number of donors, and for clerical work in conjunction with panels and at bleeding sessions if required.

The circular also spelled out its limited responsibilities in terms that could not have been clearer:

British Red Cross personnel CANNOT undertake responsibility for:—

1. The actual bleeding session.
2. Any medical work in connection with the taking of blood.
3. Any handling of bottles of blood or transporting thereof or putting into refrigerators or Banks.
4. They can take no part in the "overseeing" part of Banked blood.
5. They are not allowed to take, or have any responsibility for, Grouping or Grouping Sessions beyond calling prospective Donors and recording on the Donor's Card the result of Grouping as given them by the Laboratory or Medical Authority concerned.
6. The grouping information to be recorded on the Donors Card cannot be accepted verbally but must be in writing.[63]

Early the following year the Nairobi Red Cross cut back further, eliminating a part-time assistant for donors at the African service. Henceforth, wrote Mrs. R. Lewis to E. P. Rigby, director of medical services, "we cannot do more than supply the names of institutions and the cards of Africans who are willing to give blood and we must ask [the] Medical Department to make arrangements direct with such institutions when they wish to go out and collect blood." The Red Cross continued, however, to call up people for the European and Asian donor sessions held at the Red Cross.[64]

The British Red Cross also helped establish blood transfusion services in Tanganyika and colonies in West Africa, although they were more decentralized and began later. Tanganyika was too big to run a centralized service, but even in Dar es Salaam, early calls for a citywide service in the late 1940s and early 1950s repeatedly drew disappointing numbers of donors and transfusions. Several up-country Red Cross branches assisted local hospitals in maintaining donor panels, but reports into the late 1950s indicated that the majority of transfusion donors were found by family and friends of patients.[65]

In the Gold Coast, although there had been intermittent transfusions before the Second World War, it was the arrival of George Edington at the pathology laboratory of Accra Hospital and the subsequent hospital expansion that prompted the beginning of regular blood transfusion in the colony.[66] In 1952 he requested the Red Cross branch of the Gold Coast to begin a service, and in 1953 almost three hundred blood donations were made.[67] Donations rapidly grew not only in Accra but also Kumasi and other towns and hospital districts. By 1957 the amount of blood collected for transfusions in the newly independent country of Ghana was 2,336 pints. The Red Cross voluntary donation model was followed only in Accra, however, and even there as well as elsewhere in the country, the majority of patients brought relatives or friends to donate blood.[68]

The size of Nigeria and the decentralized nature of British rule there precluded the establishment of a national transfusion service. Generally the hospitals in the south were quicker to adopt transfusion, with Red Cross branches in Enugu, Ibadan, and Lagos responding to requests beginning in 1953 from medical officers at local hospitals for assistance in recruiting donors for blood transfusions. The Yaba laboratory in Lagos acted as a central supplier for several hospitals in the capital, until a true blood bank was established, in the late 1950s. Over two thousand pints of blood were collected in 1959 and over five thousand in 1960. But the first modern blood bank and transfusion service in Nigeria was established in the new hospital built at Ibadan and inaugurated in 1957. This six-hundred-bed facility soon reported five thousand blood transfusions annually and used all the techniques of recruitment, including propaganda, with help from the local Red Cross branch. (For more, see chapter 5.) But this was a hospital transfusion service, not an independent government or Red Cross service for the hospital. Other British colonies, such as Gambia and Nyasaland (today, Malawi), to be mentioned later, followed the same patterns as elsewhere in British East and West Africa.

Transfusions in the Belgian Congo after the Second World War

The expansion of blood transfusion in the Belgian Congo after the Second World War was unusual, primarily because the colony was so large and was served by such a variety of medical organizations. For example, the Red Cross played a key role in the process, but so too did the University of Louvain, which ran two hospitals in the colony. In addition, the Union minière du Haut Katanga (UMHK), a private company with mining concessions in Upper Katanga Province, maintained hospitals that in effect provided health service for an entire province in the south; and a big government hospital served the Congolese in Léopoldville. All these except the last are examples of hospitals with outside sources of support and expertise. As a result, these different medical facilities in the Congo had people with the knowledge and a mixture of resources that allowed them to do blood transfusions, but paradoxically that same diversity hindered the ability to organize a centralized colonial, or eventually a national, system of blood transfusion in the Congo. Of course, other circumstances were also a hindrance, such as the size and differences within the colony itself, plus the Belgian decision to severely restrict the upper levels of the educational system. The closest the Congo came to attempting such a transfusion service was in Léopoldville following the Second World War.

One key person in the transition from the interwar years to the period after 1945 was Joseph Lambillon, who began his work in one of the hospitals run by Fomulac (Fondation médicale de l'Université de Louvain au Congo), established in 1926 by members of the medical faculty at the Catholic University of Louvain. This was in response to a speech by an inspirational Jesuit missionary who visited the university in 1924. Afterward the students persuaded some faculty members to take action, and they decided to establish a hospital in 1927 at a large Jesuit mission in Kisantu, one hundred kilometers south of Léopoldville, in Lower Congo. In 1932 they established another hospital run by the White Fathers Catholic missionary order at Katana, in Kivu Province, at the eastern end of the Congo.[69]

Lambillon was specially selected and trained before he went to Katana in 1938. In fact, his posting was delayed two years for extra training in surgery and gynecology in the service of Georges Debaisieux and Rufin Schockaert at Louvain. In the meantime another doctor, N. Denisoff, was appointed to bridge the time between Lambillon's availability and the retirement of the doctor he was to replace.[70] Once there, Lambillon set out to modernize Katana Hospital, receiving funding to expand the facility from 80 to 207 beds for Africans, with support continuing even after the outbreak of the Second World War. The results of Lambillon's work were quickly evident in the article mentioned in the previous chapter that he and Denisoff published in the 1940 volume of the *Annales de la Société belge de médecine tropicale* describing the experience with blood transfusions.[71]

Lambillon remained in Katana through the end of the war and then became the director of the maternity section of the Hôpital des congolais, the largest hospital for Africans in the capital of Léopoldville. His service naturally afforded opportunities on a much larger scale for transfusion and other medical research.[72] It was in this position at the Hôpital des congolais that Lambillon came in contact with Claude Lambotte and his wife, Jeanne, in 1947, when they arrived to establish a pediatric clinic in Léopoldville for the Belgian Red Cross.

The Congo Red Cross was a branch of the Belgian Red Cross, which had first been invited to work in the Congo during the rule of King Leopold. It suspended operations in 1909 but restarted in 1924, initially at Pawa, in a rural area of Orientale Province (in the northeastern part of the Belgian Congo), focusing on the problem of leprosy.[73] Soon, however, other local medical needs of nonlepers prompted the Red Cross to establish a number of dispensaries and eventually three large hospitals, which provided general medical services. Despite growing numbers of patients including surgical (the hospital at Medje had one hundred beds and reported over three hundred major operations in 1952), there were no reports of transfusions either before or immediately after the Second World War.[74]

The Red Cross was inspired to provide medical service in Léopoldville, in part because of the need and also because there were so many Red Cross members living in the capital. Again, they initially decided to focus on treatment of a specific medical problem—in this case venereal disease—by setting up a dispensary in 1929 in Léopoldville-Est, the largest Congolese neighborhood in the capital. A new dispensary was built in 1936, and by 1945 it reported doing over sixteen thousand examinations and giving over eighty thousand injections to patients for the year, in a city whose population was estimated at eighty thousand.[75]

After the war, the Léopoldville Red Cross committee, followed the lead of the Belgian government and the Van Hoof report, which called for new resources in health to be devoted to expectant mothers and children. In 1945 the committee decided to establish a pediatric service and constructed a facility in Léopoldville, using funds from the sale of a special postage stamp at war's end.[76] Half the expense of running the service was contributed by the colonial government, with the local city government picking up 10 percent in medicine or cash subventions. Still, 800,000 francs had to be raised annually by the local committee. To direct the new center the Red Cross recruited an unusual team of two doctors: Claude Lambotte and his wife, Jeanne.

The couple arrived in the Congo, much as Lambillon had done ten years earlier, full of new ideas and innovations they were eager to bring to a place so obviously in need of them. Claude Lambotte went to medical school in Belgium, but he had trained with one of the top pediatricians in the world: Robert Debré in Paris. Claude's wife, although trained in otorhinolaryngology, joined her husband at the pediatric center and together they recruited a local staff and quickly gained the attention of the colonial health authorities, who were eager to invest in new health programs. The two doctors soon began treating anemic infants with blood transfusions, which led them to develop a supply of blood for more general transfusion purposes.[77]

Husband-and-wife teams were not unusual in early transfusion services in Africa. For example, the transfusion service of the African hospital in Dakar was first directed by Louise Navaranne, a doctor who accompanied her husband, Paul Navaranne, when he was appointed surgeon in the colonial medical service at the hospital, in 1949.[78] Similarly, Una Maclean was a medical doctor who accompanied her husband to his appointment in the colonial health service at the new University College Hospital in Ibadan when it opened, in 1957; and thanks to her previous term of service in Scotland drawing blood for a local blood bank, she became the director of the blood bank at the new Nigerian hospital.[79] Even more common was the role played by the nonmedically trained wives of doctors posted to British African colonies, many of whom headed the local Red Cross panels and transfusion services organized by local branches.

Although the Lambottes were the first to organize a transfusion service with a scope beyond their own immediate needs, transfusion had been practiced for over twenty years in the Belgian Congo. Because no large government service was organized until the 1950s, however, it is difficult to estimate the total number of transfusions done. There is strong evidence from indirect and isolated reports after the Second World War that hospital services throughout the colony increasingly used transfusions as new doctors arrived from Europe, and the government invested in newer facilities that allowed more modern medical procedures. The annual report of the hygiene service in Katanga Province for 1947, for example, noted that the surgery service in Elisabethville recorded strong activity in the quantity and quality of surgical interventions for the year. This was generally explained, the report went on, "by young doctors who show a significant propensity toward more spectacular surgery than internal medicine. All are not disposed to becoming surgeons in the future, but most do not hesitate to undertake major interventions that would otherwise be the work of qualified surgeons."[80] In 1947 there were reports of eighty-three blood group tests by the bacteriology laboratory in Elisabethville and over one hundred on Congolese in Léopoldville.[81] Blood group tests were done only for transfusions at this time, the rule of thumb being that three tests were necessary for each transfusion (one for the patient and on average two to find a compatible donor).

The Lambottes reported giving pediatric transfusions at least as early as 1949, and in the first statistical report of the pediatric clinic for 1950 there was indirect evidence of transfusions: forty-three blood group tests. This number was sixty-three in 1951 and sixty-eight in 1952 but 163 in 1954, when they also announced the beginning of a new "*service public de transfusion sanguine de la Croix-Rouge du Congo.*"[82] That same year the Red Cross committee of Stanleyville announced the start of a transfusion service, but it did not lead to a colonywide blood transfusion service in the Congo or even to a citywide one in Léopoldville.

The reason was the conflict that arose between the Lambottes and Lambillon, who had become chief of maternity services at the Hôpital des congolais in Léopoldville-Est. This was the largest hospital in the colony, with several hundred beds by 1947, when it was decided to build a new thousand-bed hospital to meet the needs of the capital. Although that goal was not achieved, in the next ten years many beds were added to the existing hospital, making it one of the largest in Africa.[83]

The differences between the Lambottes and Lambillon might have had long-term consequences, but after independence it is doubtful any broader scheme agreed upon during colonial rule would have survived. It is useful, nonetheless, to look more closely at their conflict because it reflects deeper, more fundamental conditions that shaped transfusion and health care in Congo. First, all three were difficult personalities. For example, before Lambillon went to Léopoldville, after the Second World War, he sent a note protesting the delay in determining his new appointment. He so pestered local colonial officials that the minister of health had to assure them he would work hard and concentrate as much on his duties as he had done at Katana Hospital during the war.[84] Likewise, Red Cross officials had to smooth feathers in Léopoldville shortly after the arrival of the Lambottes. As Edouard Dronsart, a Belgian Red Cross official who visited Congo in 1951, wrote in his travel notes, "I naturally took care of the minor disagreements that have arisen between Mr. and Mrs. Lambotte and the medical corps of Leopoldville. Mister and Mrs. Lambotte, with whom I spoke a long time on this question, have recognized that they have at certain times perhaps been a little harsh in their appreciation of the work of other medical services."[85]

Both Lambillon and the Lambottes were eager to bring the latest modern, scientific medicine to Africa and to report their African research findings to the rest of the scientific world. Lambillon's transfusions in remote Kivu Province were an early reflection of this, and he lobbied for expansion of maternity facilities at Hôpital des congolais as soon as he arrived and saw the unmet needs of the Congolese capital. He published findings a few years after he began work there in 1948, just as he had done in Katana. Lambillon continued to expand the obstetrics service at Léopoldville and publish the results in half a dozen articles during the 1950s. He also toured the southern United States and attended an international obstetrics conference in 1954.[86]

The Lambottes were equally eager to take advantage of the opportunities they found in Léopoldville in the late 1940s. Shortly after they arrived and set up their pediatric service in the new Centre de médecine sociale building in

TABLE 2.7. **Mothers and children hospitalized, Clinique de pédiatrie de Léopoldville, 1950–53**

Year	*Mothers*	*Children*
1950	354	474
1951	631	772
1952	621	756
1953	912	1,167

Source: "Rapport d'activité du Croix-Rouge du Congo, 1927–1959," RK Congo, box 1, CRBC Brussels.

the Quartier Saint-Jean, they began planning for facilities with more space. They were successful not only because their goals served the larger plans of the colonial administration to improve maternity and pediatric service, but also because of the popular response of African mothers to the clinic. In contrast with typical colonial practice, the Lambottes offered the latest in medical care plus a different approach to smaller details. For example, they established a card file for each patient, a common feature in Europe but not Africa. This implied follow-up care as well as giving African patients a sense of identity.[87] In addition, the Lambottes recognized the need for family and social support in African medical care, allowing family members to accompany patients and set up cooking facilities to feed them. This compelling mixture of modern and traditional elicited an immediately favorable response from Africans, as reflected in the growing number of visits to the clinic.[88]

The Lambottes continued to press for new facilities, and their success reflected both their political skills and the nature of the times, when authorities were eager to expand health facilities. They began work in the Centre de Médecine Sociale of the Red Cross in Léopoldville-Est when they arrived in 1947. Then the Lambottes were able to move into a building devoted entirely to their pediatric clinic in 1951. No sooner had they moved in, however, than they began plans for a newer and even bigger facility with the most modern equipment and more space. They were helped in this regard by a timely visit from Dronsart, director general of the Belgian Red Cross in July 1951. The impression the Lambottes made on him is evident in a letter to Claude Lambotte that Dronsart wrote shortly after his return to Belgium: "I knew theoretically the work that you had accomplished, but the visit to your service made me understand that the praise I had heard about your activity was more than merited." Lambotte took the occasion to remind him of the request for a new facility, and Dronsart promptly replied, "I have not forgotten what I promised you during my short visit at Leopoldville," adding that he had already begun discussions about the expansion with Dr. Gillet, president of the Belgian Red

Cross.[89] The Lambottes also did not shy away from helping their own cause by raising funds in the Congo for the new facility. One key event was a *tombola* (raffle), presided over by Jeanne Lambotte, the highlight of which was the auctioning of a porcelain vase donated by the queen of Belgium. The new pediatric center was inaugurated in June 1953 with one hundred beds.[90]

The transfusion service that the Lambottes organized in November 1954 grew out of the needs of their patients at the pediatric center, including mothers in childbirth and, especially, severely anemic children. To provide them with transfusions, the pediatric center trained technicians for laboratory blood testing and drawing blood. As the numbers of transfusions grew, the Red Cross and the head of the colonial health service agreed in late 1953 that the Centre de pédiatrie should organize a new "service public bénévole de transfusion sanguine." Its purpose would be to recruit donors (European as well as Congolese), do medical exams and blood group tests of donors, and draw and store blood to be used by hospitals as needed. On November 15, 1953, the new service was announced in the press, including a call for donors.[91]

Among the first to sign up were fifteen members of the Amis de la Croix-Rouge. That organization, which had started with the assistance of the Lambottes in 1951, was a means of educating as well as drawing support from the people being served by the Red Cross facility. It was unusual for the Congo in that it was run entirely by Africans. As one newspaper described it, referring to Jeanne Lambotte by her maiden name of Legrand, "for years Dr. [Jeanne] Legrand has dreamed of establishing this direct contact, this propaganda by Congolese for Congolese."[92] The idea that Africans could have their own organization was quite a shock to most other Belgians, and the executive committee of the Congo Red Cross agreed to it only with the stipulation that the president of the organization should be under the control of a delegate of the Red Cross president (Jeanne Lambotte was so designated) and that "the management of funds raised by the natives is under the effective control of a European."[93]

For their part, a large number of "evolués" responded to this rare opportunity to participate in a modern institution. The organization held regular meetings and raised funds for the relief of flooding in Luluabourg (present-day Kananga), and to purchase a van for the pediatric center to bring nursing mothers to the center for the Gouttes du lait program.[94] The Lambottes realized that the Africans in the end would decide whether they would give their own blood, so they used this organization as a way of having them be the ones who asked for it. The Amis quickly organized a special action group to publicize the need for blood donors.

Despite these innovations and the Lambottes' efforts to train their personnel for laboratory work, the blood collection service was slow to develop, mainly because of lack of resources.[95] Only one room was available at the pediatric center for examinations and drawing blood, and the personnel that the Lambottes trained had to volunteer their time after their normal work hours, when they were not on duty at the center. Donors were not paid but given a sandwich and a drink. Nonetheless, the 1954 annual report of the Congo Red Cross stated that large numbers of blood units were delivered to the Hôpital des congolais, especially to the surgery, internal medicine, and pediatrics services.[96]

The following year, the annual report indicated that because of growth of demand, the location of the transfusion service was "only temporary, and the experience of this service justifies finding a new locale specifically for it in the very near future." In early 1956, however, the executive committee of the Red Cross back in Belgium expressed great concern about the possible financial drain of the transfusion service. In response to Jeanne Lambotte's letters pressing for expansion, made more urgent, she said, by the Congolese hospital debating whether to open their own transfusion service, the committee cautioned the Lambottes not to expend more on the service until the costs and a way to pay for them could be worked out.[97] The Belgian Red Cross had a great deal of experience in this regard because it operated the national blood transfusion service back in Belgium. The goal of self-sufficiency was possible only if the government agreed to pay for the cost of blood delivered to hospitals.

The executive committee approached the colonial government about a similar agreement, and in January 1957 the governor general agreed to a charge of 2.50 francs per unit. The Red Cross predicted the transfusion service in the Congo could pay for itself in three years, and in the interim the governor-general

FIGURE 2.1. Jeanne Lambotte at Centre de transfusion sanguine du Congo. *Source:* "Du sang pour sauver les blessés," *Masolo ya Congo,* May 1, 1957.

approved 400,000 francs for the service as a temporary subsidy before achieving self-sufficiency. A new publicity campaign began, including an official statement from colonial authorities, and with the inroads already established by the Lambottes and the Amis de la Croix-Rouge, by the end of 1957, 1,600 people volunteered to be donors. From these, 1,029 were selected (861 Congolese) and a total of 345 liters of blood was collected by the service.[98]

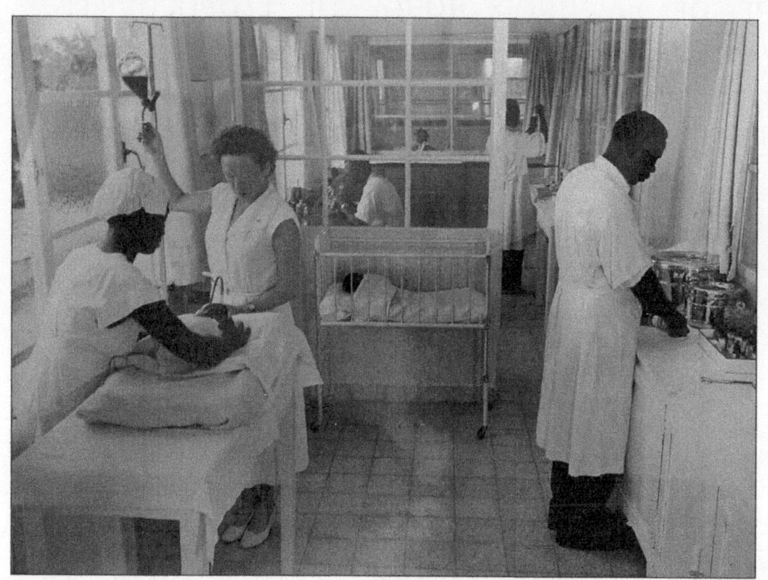

(above) FIGURE 2.2. Jeanne Lambotte giving an intrasinus transfusion to a newborn, Congo Red Cross, Léopoldville, 1953. *Source:* "Parce que les femmes noires avaient confiance en Dr. Legrand," *Moniteur belge,* June 12, 1953; also in RK Congo, foto box 32, CRBC Brussels.

(below) FIGURE 2.3. Blood donors at Léopoldville transfusion center. Source: "Voici un infirmier de ce service effectuant le prélèvement habituel sur un donneur de sang," 1959, RK Congo, foto box 32, CRBC Brussels.

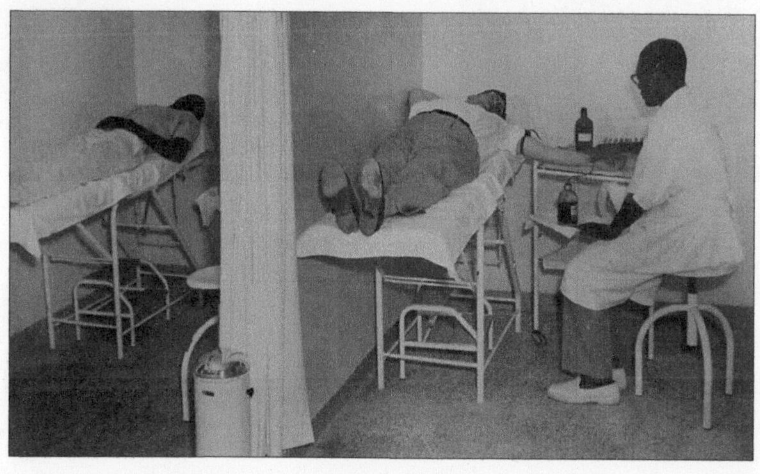

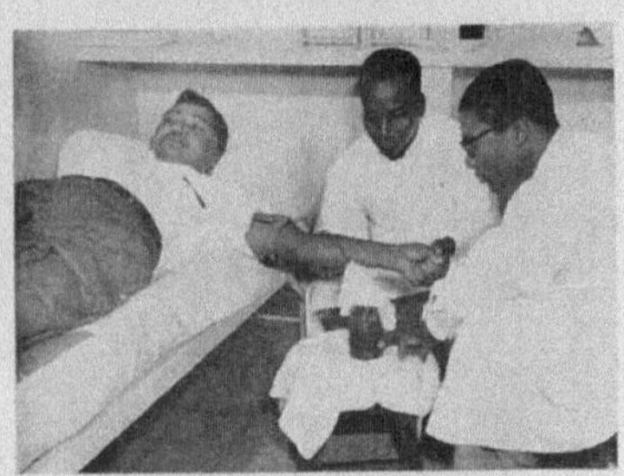

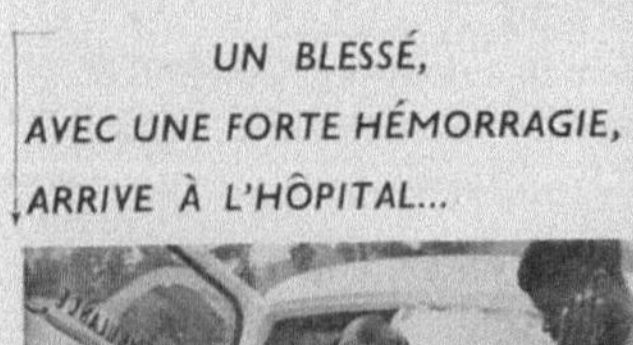

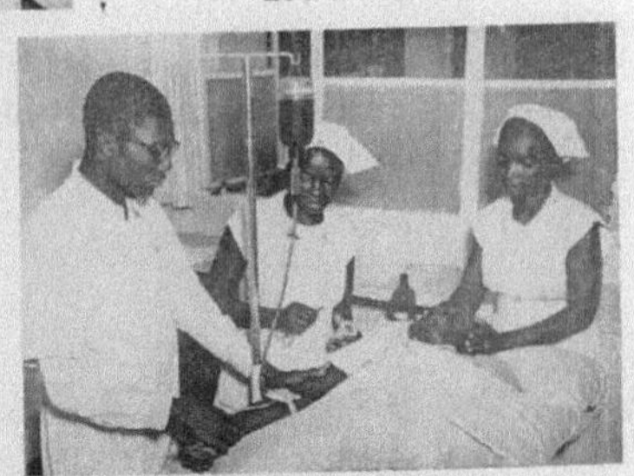

FIGURE 2.4. Poster for Centre de transfusion sanguine du Congo, 1958. *Source:* BBT-2002–24–13_a029jb, MICR.

Despite this auspicious beginning, Jeanne Lambotte (references to her increasingly used her maiden name, Legrand, at this time) soon ran into conflict, when she attempted to make the transfusion service a centralized operation. Following along the lines of European services developed between the wars which were widely emulated after the Second World War, central organizations recruited donors, collected, tested, and distributed blood on a regular basis to the hospitals, at least in the large cites. Lambillon and others at the Hôpital des congolais who did their own blood collection, would have none of that, and placed orders only with the Lambottes' service when they ran short. Jeanne Lambotte did not hesitate to criticize this practice. "Instead of placing a regular order estimating blood needs, we receive a request a few hours in advance, which is impossible to meet," she wrote in a July 1957 report.[99] The problem was that Lambotte made this criticism as part of a regular annual report with a copy sent to the Ministry of Health and the governor general of the colony.

Lambillon was livid in defending his actions, and his response was telling of the view of the *chef de service* of a large hospital. Not wanting "to worsen things," he wrote to the director of the Belgian Red Cross with copies to "Monsieur l'Inspecteur Général de l'Hygiène, Ministère des Colonies, Monsieur le Médecin provincial, Léopoldville," as well as to Lambotte. Alluding to efforts at compromise, he stated (using Lambotte's maiden name), "I do not either want to accede to a settlement made necessary only because of the abnormal and indelicate behavior of Dr. Legrand." He went on, "I will in no case agree to collaborate with an organization whose director, Dr. Legrand, has offered only sterile critiques and accusations about my chiefs of service." And he concluded, that he would drop the matter only after Lambotte had "been officially reminded of a more appropriate notion of matters and asked to exercise absolute discretion toward both me and the services that I direct."[100]

Although initially supporting Lambotte, the steering committee of the Belgian Red Cross quickly shifted its position once Lambillon's criticism emerged. The main underlying reason is that the Red Cross transfusion program paled in comparison to the needs of the very large Hôpital des congolais in Léopoldville-Est. For example, in 1956 the hospital reported 7,023 transfusions in its pediatric service alone.[101]

In addition, the Red Cross had recently begun plans to transfer its health services in Léopoldville to the Congo government to run. This had already been done with its venereal disease clinic, and the decision was also made to do so with the pediatric service, at an appropriate time. The plan, as reported in the February 1956 minutes and expressed by the president of the executive committee, was that "the medical activities of the Congo Red Cross are expected to disappear—with the exception of our facilities in [the interior province of] Nepoko, for which the government provides a 100 percent subsidy for general medical assistance—to return to activities more appropriate for a Red Cross, such as first aid and ambulance training, Red Cross Youth, etc."[102]

The matter came to a head in 1958 when the Belgian Red Cross executive committee in Brussels decided not to renew Jeanne Lambotte's contract for medical work. Despite her high visibility, she had remained an annual contract employee, even though her husband was engaged in the colonial medical service with multiyear contracts. The committee hoped that her termination would also prompt Claude Lambotte to resign, but he refused and Jeanne indicated she would fight the dismissal.[103] Tragically, before the matter was resolved, Legrand became ill and died in 1960. Her husband, Claude, remained in the newly independent Congo, taking the position of professor of pediatrics at Elisabethville until the 1970s, when he returned to Belgium to a practice in Liège.

Despite these developments, the transfusion service continued and grew, obviously complementing the large hospital service at the Hôpital des congolais. This conflict between centralized and hospital-based transfusion was played out in different forms in most African colonies and countries, with the hospitals usually keeping their independence. The numbers for the service in Léopoldville continued to grow until independence and so too did transfusion in different hospitals and regions of the Congo. There is significant evidence of the growth of blood transfusion in Léopoldville as well as other hospitals and regions of the Congo from the end of the Second World War to independence, in 1960 (see table 2.8).[104]

By the time of independence every African country had some features of a modern health infrastructure, including hospitals with transfusion services. Most of these services began within less than ten years after the Second World War, but they developed differently depending on local resources. Among the similarities was that almost all transfusion services were hospital

TABLE 2.8. Blood collection and testing, Belgian Congo, 1945–61

ar	CDP	Astrid	Hôp Cong	UMHK	Coquil	Lulua	Stanley
45		824 units coll.		1,249 tests			
46		926 units coll.		819 tests			
47		138 tests					
48				1,350 tests			
49							
50	43 tests						
51	63 tests						
52	68 tests						
53							
954	629 tests	6,490 tests		1,168 tests			
955				1,165 tests		3,678 tests	
956			7,203 transf.	1,516 tests		4,702 tests	
957	345 l. after Mar.			3,746 tests	286 tests		
958	890 l. (1,918 donations)			3,730 tests	1,087 tests	1,226 transf.	
959	2,723			3,796 tests			1,229 transf.
960	1,114 l.			2,857 tests			
961	100 l./mo.			5,565 tests			

ey:
CDP — Centre de pédiatrie (later, Service national de transfusion, Léopoldville)
Astrid — Princesse Astrid laboratoire, Léopoldville
Hôp Cong — Hôpital des congolais, Service de pédiatrie, Léopoldville
UMHK — Union minière de Haut Katanga, medical service labs
Coquil — Coquilhatville Provincial Medical Laboratory
Lulua — Luluabourg Hospital laboratory
Stanley — Stanleyville Hospital laboratory

ources: Reports cited above, both published and in CRBC Brussels archives and AA.

based, with the exception of capital cities, where a central system often served all hospitals. Uganda, Kenya, Senegal, and the Ivory Coast had some features of a national system, although for different reasons. In the case of Uganda and Kenya, this was a legacy of the Red Cross assistance in transfusion services, and in the former French colonies, it was a legacy of the French model of a national transfusion system.

The extent of transfusions was substantial in most countries, according to the numbers of transfusions reported at or around the year of independence of a number of African countries.[105] As will be seen in the next chapter, the momentum to increase the use of blood transfusion continued after the independence of new African countries in the 1960s.

TABLE 2.9. **Transfusions reported at or around independence of African countries**

Country	*Date of independence*	*Transfusions*	*Notes, date of report*
Cameroon	1960	438	Yaoundé, 1960
Chad	1960	1,467	1,421 plasma
Congo (Léopoldville)	1960	3,317	Léopoldville, 1959
		1,229	Stanleyville, 1959
Congo (Brazzaville)	1960	128	Brazzaville only
Dahomey	1960	1,325	1962
Ghana	1957	2,336	
Ivory Coast	1960	1,857	Abidjan, 1959
Kenya	1963	5,112	Nairobi
		5,392	rest of country (1962)
Malawi	1964	1,068	Blantyre
Niger	1960	167	Niamey, 1959
Nigeria	1960	5,488	Lagos
		5,770	Ibadan
		903	Northern, 1965
Senegal	1960	10,114	1961
Sierra Leone	1961	1,148	1959
Tanganyika	1964	2,400	
Uganda	1963	14,000	
Zambia	1964	10,000	

Sources: "Rapport annuel, Institut Pasteur de Cameroun, 1960," box 39, IPO-RAP, Archives of Institut Pasteur, Paris; "Territoire du Tchad, Service de santé," 1960, box 123, IMTSSA; Croix-Rouge du Congo, Section de la Croix-Rouge de Belgique, Rapport, 1959, CRBC Brussels archives; Congo, Brazzaville, Service national de la statistique, annuaire statistique, 1960, IMTSSA; Dahomey, Ministère de la santé publique, "Rapport médical sur l'activité du service de santé de la république, 1962"; Gold Coast Colony, *Report of the Ministry of Medical Research Institute,* 1957; Baba Sy, "Fonctionnement de la banque de sang de la Côte d'Ivoire," *Transfusion* (Paris) 3, no. 1 (1960): 47–51; Rogoff to Director of Medical Services, October 29, 1962, BY4/69, KNA; Malawi [Nyasaland until 1964] Red Cross, *Annual Report,* 1964, Acc 0076/30(1) Malawi, 1957–66, BRC London; "Territoire du Niger, Service de santé, 1959," box 64, IMTSSA; Federal Republic of Nigeria, *Annual Report of the Ministry of Health,* Nigeria, 1960; Northern Nigeria, *Annual Report of the Ministry of Health,* 1965; République du Sénégal, Ministère de la santé publique, Activité du service, 1969; Sierra Leone, Report on the Medical and Health Services, 1959; Republic of Tanganyika, Ministry of Health, annual report, 1964; Uganda, Ministry of Health, annual report, 1963; Republic of Zambia, *Ministry of Health Annual Report 1965.*

3 BLOOD TRANSFUSION IN INDEPENDENT AFRICAN COUNTRIES

On December 2, 1964, President Léopold Sédar Senghor of Senegal inaugurated an expansion of the Centre national de transfusion sanguine in Dakar that doubled the size of the largest blood collection facility in sub-Saharan Africa. Funding came from the Fond éuropéen de développement d'outre-mer, which had been established by the six countries of the new European Common Market to provide development assistance to countries of Africa, the Caribbean, and the Pacific. Although the transfusion center had been created in 1951 to supply blood for all French West African colonies, by 1960 its facilities were at capacity just to meet the growing demand of the newly independent country of Senegal.[1] Modernization and expansion were still needed, according to the director, Jacques Linhard, in his remarks made at the inauguration. "During the first ten months of this year [1960] we had 8,820 blood donations, whereas we need double or triple that amount. . . . Think about it, transfusions are needed from birth to death but donors are only found between the ages of eighteen to fifty-five."[2]

The expansion of the Dakar transfusion center demonstrates, in a concrete way, two important features about blood transfusion in African countries after independence: its continuing growth and the ability of countries to obtain assistance from sources besides former colonial rulers. The independence of African countries represented landmark events affecting all areas of life. In the short run for hospitals and medical care, however, there was a continuation of growth that had begun at the end of colonial rule. This was, in part, because governments wanted to demonstrate their effectiveness, but it was also because more opportunities for foreign assistance were now available to replace subsidies or investment provided by former colonial rulers. As a result, when independence came to African colonies, the practice of blood transfusion continued to grow.

Tracing that growth in countries that cover almost an entire continent is much more difficult after independence than during the colonial period. When only a few European countries exercised colonial rule, they imposed common structures and policies, especially for new practices such as modern medical care. Any similarities after independence are therefore noteworthy because many legacies of colonial rule disappeared, and similar practices likely continued because of common circumstances in different countries regardless of colonial past. Examples of forces shaping similar transfusion practice include such things as common standards of medical practice, as well as a country's level of economic development and political stability.

The first part of this chapter will offer some observations about these common features, but events in a few selected countries will also be examined in more detail to illustrate how this history unfolded. The second part of the chapter will examine changes in transfusion after the 1970s that affected all African countries, including the economic changes following the oil crisis and the rise of international organizations offering assistance and coordination of blood transfusion, especially the League of Red Cross Societies and the World Health Organization. The impact of the AIDS epidemic on transfusion will be examined in more depth in a later chapter on risk.

The First Phase of Independence: Continued Growth to the Mid-1970s

Independence came to sub-Saharan Africans at different times, beginning in 1957 for Ghana and in 1960 for most French and a few other British colonies plus the Belgian Congo. British colonies in East Africa followed a few years later, but in southern Africa it was only after extensive fighting in the 1970s that Portuguese colonies and Rhodesia became independent. For much of sub-Saharan Africa the achievement of independence did not significantly affect transfusion services after colonial rule except in circumstances of war, as in the former Belgian Congo and southern Africa. More typically, changes emerged gradually and indirectly as a result of the development of a country's health services.

Typically, at independence all countries had at least one large hospital, most often in the capital, which included modern facilities for surgery and accompanying services such as radiology, anesthesia, and blood transfusion. Beyond this, the extent that transfusion was practiced in a country depended to a large extent on the number of other hospitals outside the capital that had facilities for transfusions. In Kenya, Uganda, and Senegal, for example, colonial health infrastructures were well developed, including regional and district hospitals. In June 1962, Uganda (population est. 7.3 million) had 42 hospitals, divided evenly between government and private, with a total of 3,613 beds just before the opening of the new nine-hundred-bed Mulago Hospital in Kampala. In 1961, Senegal (population est. 3.3 million) had seven hospitals with a total of 2,556 beds (of which 1,393 were in Dakar). And in 1962, Kenya (population est. 8.6 million) reported 9,113 basic-service hospital beds in the colony, mostly outside the capital (only 1,041 in Nairobi).[3]

In most countries there were enough resources to continue this growth of hospital-based Western medicine after independence, and the expansion of blood transfusion followed accordingly. At the end of 1961, Tanganyika (population est. 10.5 million) reported 95 hospitals (half run by the government) with a total of 9,921 beds. In Nigeria at independence (population est. 43 million) there were three hundred hospitals, also about half run by the government.[4] Other countries such as the Ivory Coast and Zambia also belong in this category of places where there was continued expansion of Western medical care, including blood transfusion, after independence—from an already significant level at the time of independence.[5]

With some exceptions, blood transfusion in most sub-Saharan African countries followed a mixed model of blood collection. That model varied considerably, but typically hospitals relied primarily on their own resources to organize and recruit blood donors, while the national services provided such requirements as supplies for testing, plus some blood, depending on availability and distance. Most of the smaller and poorer countries lacked national services, but so too did places like Nigeria and Zaire, albeit for a different reason. Initially this was because of their large size and complexity but also because immediately after independence, there was political instability and civil war. Even during periods of peace and stability, neither of these two large countries of sub-Saharan Africa had been able to organize colonywide or national transfusion services before or after independence. According to available reports and publications, despite (or perhaps because of) lack of centralized transfusion services, the numbers of transfusions continued to expand in these countries as well.

There is ample evidence that this overall increase in the use of transfusions continued in Africa. Taking some representative reports of transfusions in a dozen different countries (including many for large city hospitals), one can see the growth by the 1970s compared to the previous decade.[6]

Every country reported increases, although with some noteworthy differences. For example, the increase was the greatest in Yaoundé, Brazzaville, and Kampala thanks to the creation of new blood banks and hospitals around the time of independence. Other places showing less dramatic growth had experienced increases in blood collection from earlier investments in new facilities. This was the case in Senegal (see chapter 2), although Linhard's hope for

TABLE 3.1. Growth of blood transfusion, selected African locations, 1960s–70s

Location	*Transfusions*	*Date*	*Transfusions at or near independence*
Benin (Cotonou)	2,202	1972	1,325 (1962)
Cameroon (Yaoundé)	4,543	1972	528 (1960)
Cameroon (Douala)	1,825	1972	—
Chad	2,376	1976	1,467 (1960)
Congo (Brazzaville)	6,103	1972	128 (1960)
Kenya (Nairobi)	16,324	1973	5,112 (1963)
Nigeria (Ibadan)	8,773	1973	5,770 (1960)
Senegal (Dakar, CNTS)	11,695	1970	10,114 (1961)
Sierra Leone	2,049	1970	1,148 (1959)
Tanzania (Dar es Salaam)	2,500	1965	907 (1960)
Uganda	34,311	1970	14,000 (1963)
Zaire (Kinshasa)	5,702	1972	3,317 (1960)
Zambia (Lusaka)	6,993	1971	
(Ndola)	5,713	1971	

Sources: Ayité, "Transfusion sanguine; République du Tchad, Ministère de la santé publique, *Annuaire de statistique sanitaire,* 1976, 82; Kenya, Medical Department, *Annual Report 1966,* 154; Sierra Leone, *Report on the Medical and Health Service,* 1959, 1970; Institut Pasteur du Cameroun, *Rapport,* 1960–62, 69.

increased blood donations were never realized. Likewise, in Nigeria, University College Hospital in Ibadan provided a dramatic increase in the ability to collect blood for transfusion when it was opened, in 1957.

Funding for these improvements in facilities was greatly assisted by increased external opportunities for African countries in the 1960s and early 1970s. The main source was the richer countries that competed for the attention, potential trade, and UN votes of newly independent African countries, a process accentuated by the cold war. In addition, with Western Europe and Japan recovered and booming economically after the Second World War, attention shifted on the international scene to economic development, which also motivated the richer countries to invest in the newly emerging third world.[7] The net result, at least until the oil crisis and economic downturn of the 1970s, was the ability of the new African countries to make bilateral arrangements for economic development assistance, including health infrastructure, hospitals, and sometimes transfusion services. African countries probably had an advantage in attracting outside foreign assistance to organize their blood transfusion services because their needs were limited, and feasible projects could be developed that did not require large amounts of funding. The economic difficulties after the 1970s increased the importance of outside assistance. In addition, this coincided with new efforts of the League of Red Cross Societies and the World Health Organization to help find support, provide expertise, and respond to newly emerging diseases such as hepatitis B and AIDS that threatened the safety of the blood supply.

Blood Transfusion in the First Years of Independence: The Example of Kenya

It is impossible for this survey to present extensive analysis of the history of blood transfusion in every African country after independence. Three have been selected for closer examination because they had well-established transfusion programs at the time of independence that continued to grow, at least until the 1970s. Kenya, Uganda, and Senegal also have differences that make for useful tests of common influences that affected transfusion services regardless of francophone or anglophone traditions, settler legacies, or location in East or West Africa. The most striking similarity was that all three countries developed the beginnings of national blood transfusion systems, with facilities that served hospitals in the capital as well as some service to the rest of the country. In these outlying regions hospitals sometimes received blood from the national service, but they were on their own to recruit donors and store blood or make other arrangements (usually family or replacement donations) for transfusion. And even in the capitals, in times of shortage, the hospitals developed backup methods of collecting blood from patients' relatives, friends, or paid donors.

Whether centralized or not, blood donation and transfusions in all three countries continued to grow after independence. The records for Kenya and Uganda from the mid-1960s to the mid-1980s are incomplete, but there is sufficient evidence not only to show that growth but also to permit an estimate of the level of transfusion activity in different regions of the country. Throughout this study, only reported figures have been used to describe the extent of transfusion activity. The reason is to give a firm basis for this historical analysis. Since these reports are incomplete, however, they most likely underrepresent the real extent of transfusion use.

It is perhaps most accurate to say that Kenya was able to develop a limited national blood transfusion service. This followed from significant growth in health and hospital infrastructure that continued after independence. In 1965 the national government of Jomo Kenyatta made a major commitment to "provide medical and hospital services for all who needed it without charge."[8] Thanks to relative prosperity, political stability, and outside assistance, the success in the following decades could be seen most concretely in the growth of the numbers of hospitals and trained doctors, keys also to the practice of blood transfusion.

TABLE 3.2. **Hospital beds (government and nongovernment), Kenya, 1962–78**

	1962	*1970*	*1973*	*1978*
Hospital beds	10,617	14,537	18,186	26,922

Source: Ogot and Ochieng', *Decolonization and Independence,* 134.

TABLE 3.3. **Doctors and dentists, Kenya, 1963–78**

	1963	*1966*	*1974*	*1978*
Doctors and dentists	948	910	1,242	1,596

Source: Ogot and Ochieng', *Decolonization and Independence,* 134.

The need for training African health providers was recognized in Kenya as early 1925, but even after the creation of the Central Training Depot at the African hospital in Nairobi, in 1929, only a very limited number of hospital staff were trained.[9] A medical school was established at the University of Nairobi only in 1967, much later than the University of Ibadan Medical School in Nigeria (1948) and Makerere Medical School in Uganda (1949). It had an initial class of thirty students in a five-year course, but in 1972 the Nairobi admissions were increased to one hundred students, and as a result the number of doctors in Kenya increased dramatically. By 1993 there were approximately thirty-seven hundred doctors in the country.[10]

The growth of government health infrastructure even before independence both increased the use of transfusion and changed its operations. Although the Red Cross had been instrumental in blood collection as early as the 1940s, by the time of independence, in 1964, Red Cross activity was largely confined to the recruitment of donors. Moreover, after 1965 even that limited role was dropped except in a few regions such as Mombasa.[11] What replaced the Red Cross was largely a hospital-based system except for Nairobi, where a collection service was established in the National Public Health Laboratory at the former Native Civil Hospital (renamed King George VI Hospital in 1952, then Kenyatta National Hospital in 1963).[12] The Kenya Blood Transfusion Service collected blood mainly for the big government hospital and others in the capital and also attempted to provide regional hospitals with supplies. Government blood transfusion executive officers were appointed in some of the provinces (three in 1972, according to a WHO consultant) to organize services in their regions, give technical advice, and sometimes act as resources for regional and district hospitals. They were also responsible for reporting statistics.[13]

The country's blood transfusion service also played a key role in promoting an institution after independence that used blood donation to build a sense of national identity: Kenyatta Day (October 20). This event was begun under a different name in 1968 by Jomo Kenyatta and continued by his successor, Daniel Arap Moi, who established its new name and dedicated the event "to the remembrance of not only the founding father of the nation but also the many unheralded heroes who shed their blood for the independence and self-determination" of Kenya.[14] Each October the founder of the country was celebrated, and for a week or more before Kenyatta Day the people were encouraged to donate blood.[15]

This event was more important for public relations purposes than insuring adequate blood collection, since blood was needed all year round (see chapter 5). The establishment and growth of importance of this institution reflects another defining feature of Kenyan postindependence history: its political stability, which allowed continuity in health service including blood transfusion into the 1970s and beyond. In contrast, neighboring Uganda sank into the political and military chaos of dictatorship and civil war during this period which, of course, had repercussions for all facets of life, including health services.

Transfusion records available for Kenya from 1959 to 2000 are most complete for the Nairobi region. Intermittent reports from the seven regions of the country, nevertheless, afford an opportunity to draw some conclusions not only about change over time but differences between regions.[16]

A few comments should be made about the sources of some of these statistics. The earliest reports for the provinces were collected by M. G. Rogoff, the new director of the National Public Health Laboratory and Nairobi transfusion service, when he made his first tour of Kenya, in 1962. His figures were obviously incomplete. For example, he received a figure for Kisumu, the capital of Nyanza Province, that was only one-third of the total for the whole province reported in the annual medical department report three years later, in 1965. Another unusual source found beginning in 1979 was an attempt for a few years to compile countrywide data on blood collected. These reports show a total figure for the whole country of over 113,000 one-pint units collected by 1981.

But caution should be used in judging the reliability of any incomplete statistics, especially where there is little or no way to verify their accuracy. Nonetheless, without being overly naive, one can find enough consistency in reports from different places over time to draw a few conclusions about the practice of blood transfusion they describe. For example, the most consistent records of blood donation were in Nairobi, where the regularity of reports, plus the fact that they were produced in the capital and done by the national

TABLE 3.4. **Blood donations, Kenya, by province, 1959–88 (pints)**

Year	*Total*	*Nairobi (BTS)*	*Eastern*	*North Eastern*	*Western*	*Central*	*Rift Valley*	*Nyanza*	*Coast*
Earliest report of transfusion service (when known)		*(1947)*					*(1953)*	*(1955)*	*(1952)*
1959		5,146							
1960								953	
1961								1,772	
1962	11,282	5,890	656			1,595	845	654	1,642
1963		5,112						1,255	
1964		5,612						1,687	
1965		7,711						1,865	
1966		10,020							
1967		11,729							
1968		14,053							
1969		14,295							
1970		14,644							
1971		16,490							
1972		15,117							
1973		16,324							
1974		15,704							
1975		16,561							
1976		18,070							
1977		18,288							
1978		16,725							
1979	86,777	21,106							
1980	103,512	23,513							
1981	113,189	26,684	10,810	703	8,837	14,912	16,233	19,488	14,326
1982	113,706	20,831	13,840					16,068	
1983	96,972	20,231	12,187	3,820	9,946			21,036	
1984		18,845	11,209		13,205	10,474	16,730	23,148	
1985		19,729			17,594	18,171		28,395	
1986		16,798							
1987		17,477			12,738		23,506		9,145
1988	83,162	14,216	4,824			10,312	19,390	18,420	8,669

Sources: Kenya, Colony and Protectorate, *Ministry of Health Annual Report 1959;* Republic of Kenya, *Ministry of Health Annual Report 1965.*"See also Republic of Kenya, "Country-wide annual report—Blood donation—1981," June 11, 1982; Kenya Ministry of Health, "Memorandum on 1984 Kenyatta Day Celebration (32nd)," October 3, 1984; Republic of Kenya, Blood Donation Annual Reports from provinces and Nairobi, 1986–88; Blood Donor Service Nairobi, "Country-wide Blood Donation Report for the Year 1988," February 8, 1990, KNBTS; Rogoff to Director of Medical Services, October 29, 1962, BY4/69, KNA.

government's laboratory, adds to their credibility. For twenty years, from 1962 through 1981, there was an impressive five- to sixfold increase in blood collected for Nairobi. Yet the incomplete reports from the outlying provinces suggest an even greater growth in the use of transfusion there. Thus, Rogoff found in 1962 that there was already at least as much blood collected in the provinces of Kenya as was collected in Nairobi. By 1981, however, there was three to four times as much blood collected in the provinces as in Nairobi. Put another way, there was as much as a tenfold increase in blood collection for transfusion in the provinces of Kenya outside Nairobi, using Rogoff's figures as a starting point.

Because there is some regional data from Kenya, it is possible to extrapolate in order to estimate the ongoing overall level of transfusion activity for the country as a whole. The assumption is that if a region reported data for one year and not the next few years, it was not because transfusion ceased but rather that data was not recorded or reported. When the statistics reported for Nairobi and the extrapolated statistics for the rest of the country are plotted on a graph, the dramatic difference in the growth of transfusion can be seen much more clearly. By the 1980s blood donations reported outside Nairobi were as much as five times the amount reported in the capital.

A final feature of the record of blood donations in Kenya was a decline beginning in the early 1980s, which is of special note because the timing suggests

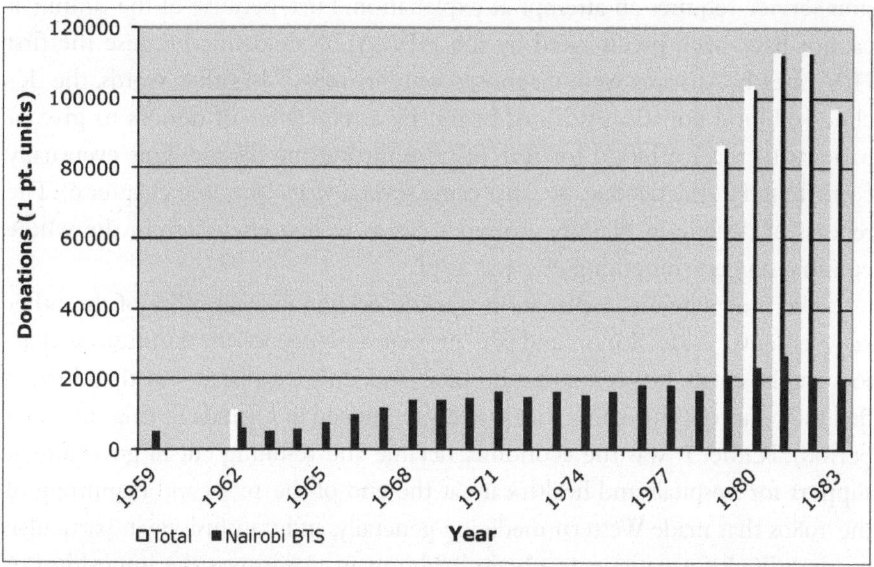

FIGURE 3.1. Kenya blood donation, 1959–83. *Source:* Blood Donation Annual Reports from Provinces and Nairobi and "Country-wide Annual Report—Blood Donation," various, 1959–83, KNBTS; Rogoff to Director of Medical Services, October 29, 1962, BY4/69, KNA.

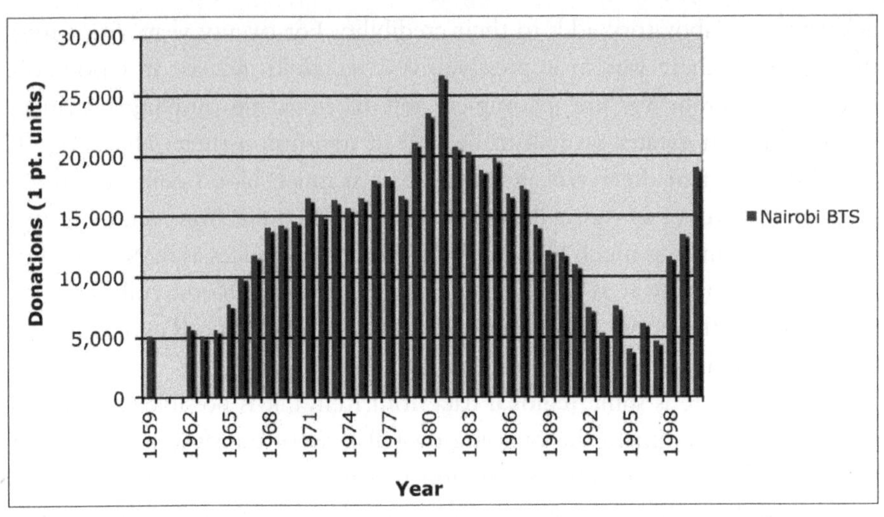

FIGURE 3.2. Nairobi Blood Transfusion Service donations, 1959–2000. *Source:* See fig. 3.1.

that the drop began before the appearance of HIV/AIDS in Africa. The figures reported for Nairobi from 1969 through 2000 show a steady rise from around five thousand donations in 1962 to a peak in 1981 of over twenty-six thousand. Then reports show a decline back to around five to seven thousand in the mid-1990s and a dramatic rebound to almost nineteen thousand in 2000.[17]

This decline in reports of blood donations at the Nairobi blood transfusion service requires an attempt at explanation. First, because of the timing, it cannot have been precipitated by the HIV/AIDS epidemic because the first HIV cases in Africans were diagnosed only in 1983.[18] In other words, the decline in blood donation was not begun by a reluctance of donors to give or patients to receive blood for fear of transmitting the disease. This eventually contributed to the decline, but that came several years later (see chapter 6). The reports of decline in Nairobi around 1980, as well as elsewhere in the whole country, suggest something else was at play.

The most obvious explanation was the decline in availability of the other key elements besides donors and patients that were needed for transfusions: doctors and hospitals. It was not that hospitals suddenly disappeared or that doctors fled the country (something that actually happened in Uganda during this same period). Rather it was the economic decline and resulting cut in government support for hospitals and health care at the end of the 1970s and beginning of the 1980s that made Western medicine generally, and transfusions in particular, more difficult for patients to obtain.[19] The main reason was the imposition of user fees at hospitals, but in addition there were shortages in supplies and a lack of transportation to recruit donors. Reports from the provinces and the Nairobi transfusion service contain many complaints of shortages that continued through the end of the decade. For example, a 1988 report from the Rift Valley Provincial General Hospital explained a decline of over four thousand units of donated blood from the previous year as follows, "The drop is mainly attributed to the shortage of blood bags [and] lack of funds and transport to take the blood donor units to the places where we had made appointments."[20]

The decline in blood donation was not reported everywhere in Kenya. Nairobi showed by far the biggest decline, but this may have been the result of there being one major institution that used blood: Kenyatta National Hospital. By 1982 it had grown to two thousand beds, employed over half the government doctors, and received 40 percent of the Ministry of Health's recurrent budget.[21] But as the figures for the rest of the country show, other regions varied greatly in blood collected between 1981, the peak of donations, and 1988, when complete statistics were reported again from all regions.[22] (See table 3.4.)

Although some of the intermittent reports from smaller regions suggest declines similar to Nairobi (e.g., Eastern and Coast), the large regions of Western, Rift Valley, and Nyanza held steady. One possible explanation is that there was underreporting of blood collection. Since Kenya's was largely a hospital-based transfusion system, any hospital could draw blood from a relative or friend (real or paid) to be transfused immediately to a patient. This served as a safety valve in emergency situations, and it is possible that hospitals, especially in the provinces, used this source of blood when there were shortages and did not report it.

The precipitous decline in reports of blood collection in Nairobi beginning in the mid-1980s is clear; but it seems unlikely that this was entirely the result of a drop in the number of transfusions in such a growing city (Nairobi's population in 1979 was 828,000; by 1995 it was 1.8 million). Yet reports of blood donation declined from 25,000 in 1982 to less than 4,000, only to rebound in 2000 to 19,000 when outside funding began to help modernize the transfusion system.[23] It is possible that hospitals in the city, including the huge Kenyatta National Hospital, also collected blood on-site from relatives and friends of patients,[24] but the same budget cuts that brought chaos in the hospital also ended systematic reporting of statistics for blood transfused.[25]

Even if there was a decline, Kenya had a high level of blood transfusions that continued into the early 1980s, which meant, among other things, that it unfortunately contributed to the spread of HIV/AIDS before the epidemic was discovered. The country acted quickly, however, to establish HIV testing of the blood supply. Not suffering the internal disruption that affected Uganda, Kenya's hospitals were ordered to start screening for HIV by a June 1985 memorandum from the Kenyan Ministry of Health, beginning with all blood donors in Nairobi.[26] By 1987, the Eliza test for HIV was distributed to most regional hospitals and in the next few years to district hospitals collecting blood for transfusion.[27]

Blood Transfusion in the First Phase of Independence: The Example of Uganda

Although no sub-Saharan African country had a truly centralized transfusion system, Uganda came closest because of its history, resources, and geography.[28] Late in 1960 the Uganda government established a national transfusion service, appointing Jean T. Holland as director. This development effectively fulfilled the prediction mentioned in the previous chapter, when the colonial governor of the protectorate declared to the annual meeting of the Uganda Red Cross in 1952 that blood transfusion "eventually no doubt will become the responsibility either of local government or of Central Government."[29] The national service helped the practice of blood transfusion continue its growth after independence in Uganda, as in Kenya, because of the growth in health infrastructure. Thus the number of hospitals (governmental and private) rose from fifty in 1965 to seventy-six in 1980, but that growth was slower than in Kenya and failed to keep up with the population, which grew from 7,551,000 to 12,636,000.[30]

The Uganda Red Cross remained involved after the new government transfusion service began; in fact, it received a subsidy for donor recruitment. Supplies and materials (bottles, needles, tubing, etc.) were provided by the National Transfusion Service, which moved into the old European hospital on Nakasero Hill, after the new hospital at Mulago opened in 1962.[31] In reality, the national service collected blood only for the hospitals in the capital, Kampala. Out in the provinces and districts, not to mention the private missionary hospitals, blood was collected by each institution. There were similarities in donors—older school children, government employees, the army, police trainees, and prisoners—and the national service provided supplies and technical expertise, including work with local Red Cross chapters to assist in recruitment with films and general publicity. One reflection of the limits of the national service is the lack of consistent or complete records reporting blood collected or transfusions done.[32]

The history of blood transfusion in Uganda shows a pattern of growth in the 1960s and 1970s that was similar to other African countries such as Kenya, but with a decline at an earlier time. This was because of internal political and military problems associated with Idi Amin's regime and the civil war that followed, in the late 1970s and 1980s. The overall figures are more complete for the earlier period but not always available by province. Nevertheless, these figures are similar to Kenya, showing a growth of transfusion in the capital but an even faster increase in the rest of the country.[33]

As in Kenya, the most complete early figures for Uganda are from the capital region of Kampala. Provincial data are much less complete, but there are figures in the early 1970s from Red Cross reports of district blood collection campaigns around the country. These figures were not complete, hence they do not represent the total amount of blood collected, but they include the main regional centers, such as Gulu, Jinja, Mbale, and Soroti. The figures given are consistent with those reported for these locations in the 1960s, so there is

TABLE 3.5. **Blood collected, Uganda, 1956–2005 (pints)**

Year	*Nakasero (Kampala)*	*All Uganda*	*Year*	*Nakasero (Kampala)*	*All Uganda*
1956	1,236	1,407	1986	6,300 (est.)	—
1957	1,183	1,409	1987	6,300 (est.)	—
1958	2,057	2,726	1988	—	—
1959	2,530	3,875	1989	—	7,000
1960	—	—	1990	—	13,500
1961	—	8,553	1991	—	19,000
1962	5,960	9,000	1992	—	28,000
1963	—	14,000	1993	—	30,000
1964	—	—	1994	13,551	37,967
1965	11,500	17,583	1995	14,191	38,854
1966	13,129	—	1996	15,816	43,022
1967	13,682	—	1997	—	48,860
1968	14,103	—	1998	—	67,230
1969	14,307	—	1999	—	68,493
1970	15,721	—	2000	—	81,000
1971	15,288	—	2001	—	86,805
1972	17,064	38,071	2002	—	92,247
1973	11,835	34,311	2003	41,312	102,703
1974	14,476	33,855	2004	—	106,996
1975	11,760	24,227	2005	—	115,000
1976	9,503	22,007			
1977	9,342	21,964			
1978	8,967	23,944			

Sources: Uganda Red Cross Annual Reports, 1956–1978; Uganda Central Council Red Cross Branch, annual reports, 1956–63, Acc 0076/58(1) Uganda, 1949–1956, and Acc 0076/58(2) Uganda, BRCS archives; UBTS proposal to PEPFAR, 2004; UBTS PEPFAR report, April 2006, UBTS.

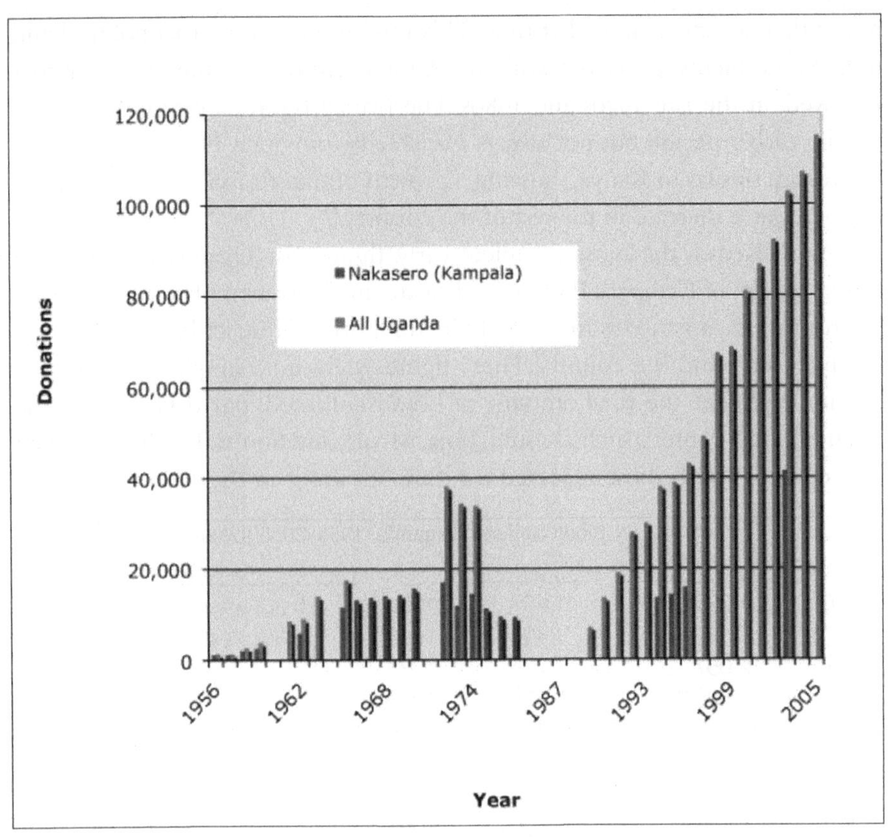

FIGURE 3.3. Uganda blood donations, 1956–2005. *Sources: Uganda Red Cross Annual Reports,* 1956–78; Uganda Central Council Red Cross Branch, annual reports, 1956–63, Acc 0076/58(1) Uganda, 1949–56, and Acc 0076/58(2) Uganda, BRCS archives; UBTS proposal to PEPFAR, 2004; UBTS PEPFAR report, April 2006, UBTS.

reason to take them as accurate. Most important, they show that by the early 1970s more blood was collected in the provinces than in the capital, in fact almost double.

The decline in blood donations in the 1970s is much simpler to explain in Uganda than in Kenya or Senegal. In addition to economic decline, the Amin regime and subsequent civil war had disastrous consequences in all areas of life, including access to health care. One crucial example is what happened to the availability of doctors. There were 450 doctors at independence, although only ninety Ugandans since it took time for the Makerere Medical School (begun in the late 1940s) to have an impact. But with a hundred students accepted each year, by 1968 there were 978 doctors in the country, the majority Ugandan. The situation deteriorated rapidly in the 1970s, first with fewer medical

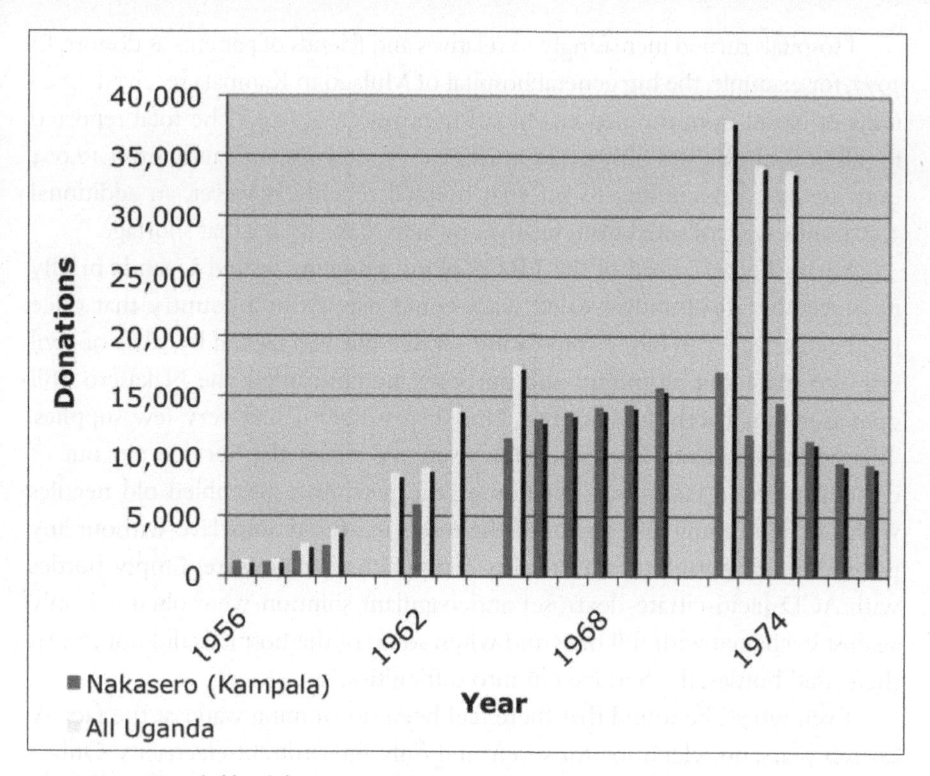

FIGURE 3.4. Uganda blood donations, 1956–77. *Sources: Uganda Red Cross Annual Reports,* 1956–78; Uganda Central Council, Red Cross Branch, annual reports, 1956–63, Acc 0076/58(1) Uganda, 1949–56, and Acc 0076/58(2) Uganda, BRC London.

students admitted, followed by students leaving school and doctors leaving the country. In 1979 there were 574 doctors in the country; from 1981 to 1984 only half the medical school graduates passed their exams; and of fifty-three who passed in 1983, twenty-four left the country.[34]

There is only scattered and personal testimony about the decline in blood collection in the late 1970s and early 1980s in Uganda, but these reports provide some clues as to what happened when blood supply could not keep up with demand. For example, in 1979, Evelyn von Steffens of the League of Red Cross Societies visited Uganda, stopping first at the Nakasero Hill blood transfusion service in Kampala. She reported, "The service has not collected blood inside the premises for years but all collections come from mobile units. Excluding the blood taken in Mulago Hospital, the figure for July–September is 638 collections." That projects to an annual rate of approximately twenty-five hundred donations, compared to around fifteen thousand donations in the late 1960s and early 1970s.[35]

Hospitals turned increasingly to relatives and friends of patients as donors. In 1977, for example, the big general hospital of Mulago in Kampala received 5,849 units of blood from the national blood transfusion service. (The total reported for all of Kampala was only 9,342 units in 1977, down significantly from 17,064 units in 1972.) According to internal hospital records, however, an additional 3,493 units was collected from relatives to help make up for that shortage.[36]

Juhani Leikola, head of the LRCS blood program, visited Uganda briefly in November 1984 and revealed what could happen in a country that once had been a leader in blood transfusion service but was racked by years of civil war and strife. For all intents and purposes, he concluded, the Nakasero Hill operation was barely functioning. Not surprisingly it had very few supplies. "Plastic bags had not been used for years and when the Service ran out of disposable blood taking sets for bottles, the personnel assembled old needles with silicone tubing and sterilized these sets in an old autoclave without any possibility of controlling the sterility, temperature or pressure. Empty bottles with ACD [acid-citrate-dextrose] anticoagulant solution were obtained only against exchange with full ones and when some of the hospitals did not return their used bottles, the Service ran into difficulties."[37]

Even worse, he found that there had been no running water at the facility for ten years, no telephone for seven, and only intermittent electricity. Only a standby generator kept one big and one small refrigerator running for blood storage. This shifted the responsibilities for blood collection largely back to the hospitals. After his visit to Mulago Hospital, Leikola noted, "during the three months between July and September 1984, the hospital had used 1,728 bottles of blood [about seven thousand annually], of which 9.5% had come from [the] Nakasero Service, the rest having been collected from patients' relatives."[38] He found a similar situation at Nysambya Hospital, the second largest in the country, with 310 beds: "In 1983 the hospital had collected 2,346 bottles of blood." The figures were not ready for 1984, but Leikola was told there had been 198 donations in September, with fifteen units received from Nakasero.

The annual Uganda Red Cross report for 1982–83 indicated that the organization continued to help with transfusions, but lamented the state of this traditional Red Cross service:

> It is sad to report that the Blood Transfusion Service programme[,] which is one of the central activities of the Society[,] has still not shown improvement. The programme is still affected by the same difficulties which were reported last year, i.e., lack of transport, equipment and refreshments as well as apathy on the part of donors. The doctor in charge of the Blood Transfusion Service also resigned during the year.

Since the situation has been deteriorating, attempts have been made to set up bleeding sessions and we have appealed to Sister Societies to assist in the rehabilitation.[39]

The only local Uganda Red Cross division to mention blood transfusion (Kampala East) stated that "individual members were encouraged to donate blood since it was considered unnecessary to carry out a massive campaign due to lack of storage facilities." Nonetheless, for 1984 through 1986 the Red Cross reported that a number of divisions continued blood collection activity.[40]

The last estimate of transfusions before the service was reorganized at the end of the 1980s came from the report of a consultant for the European Union who eventually recommended funding for a new Ugandan transfusion service.[41] In 1987, after the end of civil war, the new Ugandan president, Yoweri Museveni, invited potential donors to Kampala to discuss assistance in rebuilding the country. Unlike some other African leaders at the time, he recognized and asked for help with the HIV/AIDS epidemic.[42] Because of the critical danger of infection through contaminated blood transfusions, in May 1987 the EU sent a consultant, Dr. Lieve Fransen from the Institute of Tropical Medicine in Antwerp, to survey transfusion in Kampala. She estimated a seropositive rate for HIV at between 15 and 20 percent and collected figures for transfusions from May 1986 through April 1987, estimating that there were slightly over sixty-three hundred transfusions for the year. This confirmed the great danger of the contaminated blood supply; the only good news was that the number of transfusions reported for Kampala had not been this small since the end of colonial rule in 1962.[43]

To appreciate this decline and make a more accurate comparison of the extent of transfusions practiced in Kenya and Uganda at their peak, it is necessary to account for the different population size. Using the donation figures as an indicator of transfusion rates (roughly 90 percent of blood was used), one finds the peak years for Kenya in the early 1980s and the peak for Uganda in the mid-1970s.

TABLE 3.6. **Blood donation rates, Kenya and Uganda, 1972–2003**

Country	*Year*	*Donations*	*Population (millions)*	*Donations/100,000*
Kenya	1981	113,189	16 m (1980)	707
	2001	101,212	30 m (2002)	316
Uganda	1972–74 (avg.)	35,412	11 m (1965)	443
	1995	38,854	19 m	177
	2003	102,703	25 m	395

Sources: www.populstat.info; sources of blood donations are from data in tables 3.4 and 3.5.

Adjusting for the different population size, at least two conclusions can be drawn from the annual donation rates. First, the levels of blood donation per capita reached in the mid-1970s in Uganda were higher than thirty years later. By the mid-1990s, with a population estimated at 22 million (1995) and after the new blood transfusion service was in place, there were 38,854 annual donations, or a rate of 177 per hundred thousand. Even after eight more years of expansion and external funding, the Uganda transfusion service provided 102,703 units of blood in 2003 for an estimated 26 million people, or 395 per hundred thousand, which was less than the 443 per hundred thousand reported in the mid-1970s. This shows that current donation rates have not exceeded previous per capita rates. Equally as important, the extent of donations and transfusions at this early date suggests an underestimation of the role of transfusion in the spread of HIV/AIDS in Uganda. The findings for 1987 of 20 percent prevalence in Uganda are likely to have been significantly the result of the extensive use of transfusions and the lack of HIV testing before then. Otherwise a lifesaving measure to be sure, the widespread use (or overuse) of transfusions in Uganda also may tragically have helped spread the unknown disease more than in other countries with less developed transfusion services.

Blood Transfusion in the First Phase of Independence: The Example of Senegal

The independent country of Senegal was unusual in having benefited from earlier investments made in Dakar while that city served as the capital of the French West African colonial federation from its creation in 1902 until 1958, when it was replaced by the short-lived Communauté française. Thus, at independence the country had one of the earliest and largest transfusion facilities in all of sub-Saharan Africa, originally designed to serve a region far larger than the country itself. Although after independence Senegal, like Kenya, enjoyed relative peace and stability, its small population and lack of natural resources (in the late 1960s Senegal's population was 3.6 million, compared to Uganda's 8 million in 1971 and Kenya's over 10 million), resulted in slower expansion of medical infrastructure compared to other African countries after independence.

In 1961 the Ministry of Health of Senegal reported the country had seven major or regional hospitals with 2,556 beds. In 1969 it reported an additional hospital and the expansion of beds to 3,571.[44] The Dakar École de médecine de l'Afrique Occidentale Française, which had supplied a subclass of doctors plus nurses and midwives for much of French colonial Africa since 1921, was integrated into the new University of Dakar in 1957 (subsequently renamed Université Cheikh Anta Diop in 1987). It was only at independence, in 1960, that a new medical school recognized by European standards was begun, and it relied heavily on French instructors and admitted a large number of students from France, thus putting Senegalese students at a disadvantage for entrance and retention. As a result, in 1968, as the new medical school following stricter international standards began producing graduates, there were 214 doctors in Senegal. Of these 117 were Senegalese and most of the total (147) resided in Dakar.[45]

As in Uganda, the first director of the Senegalese blood transfusion service was a European, Jacques Linhard, who had participated in the establishment of the service at the end of the Second World War. With a few initial short periods of interruption, he held that position until 1978. Linhard also held a post at the medical school in Dakar, where he helped train the country's next generation of hematologists. As in Kenya and Uganda, the National Blood Transfusion Service of Senegal (Centre national de transfusion sanguine, CNTS), collected blood primarily from donors in the capital city and supplied it to hospitals and clinics in the region plus some outlying hospitals. The hospitals outside the capital—plus the Hôpital Principal in Dakar (the former military hospital that also served Europeans and later the Senegalese who could afford it)—also collected blood for their own use. The shorter distances and historical tradition of Senegal, however, meant that the country came close to having a true national transfusion service.

This was even more unusual because Senegal had nothing like the economic resources of Kenya or even Uganda; so after independence it was hard pressed to meet growing demands for transfusion. On the other hand, the transfusion center in Dakar had been established early (1951) and had multicolony responsibilities until independence. The reduction in geographical responsibility was balanced by the possibility of attracting other support now that the country was independent. For example, as mentioned at the beginning of this chapter, in 1964 the European Common Market helped fund the costs of the CNTS of Senegal to expand and modernize its facilities.[46] This was an early indication of one of the advantages that independent countries of Africa enjoyed that had not been available during colonial rule: alternative sources of outside funding for development projects.

After 1964, Senegal continued to collect blood and give transfusions that approached the high levels established before independence. The capacity of the CNTS had originally been developed to supply blood to all French colonies in West Africa, but even after the independence of these countries and the establishment of their own national transfusion services, the growth of demand in Dakar and the regional hospitals of Senegal warranted the expansion of the CNTS in 1964. Figures for the number of units of blood or plasma reported

TABLE 3.7. **CNTS activity, Senegal, 1960–69**

Year	*Donors*	*Units shipped (total)*	*Whole blood*	*Plasma*
1960	10,311	—	—	—
1961	10,114	10,073	8,486	1,587
1962	10,900	10,396	8,205	2,191
1963	8,853	9,621	7,728	1,893
1964	10,573	11,317	9,550	1,767
1965	13,490	12,280	10,510	1,770
1966	14,211	15,591	13,829	1,762
1967	9,945	16,142	15,203	939
1968	14,662	17,036	16,617	419
1969	11,748	13,422	13,002	420

Sources: CNTS du Senegal, "Activités pendant l'année 1961"; République du Sénégal, Ministère de la santé publique, *Statistiques sanitaires, 1961–69,* CNTS Dakar.

shipped by the CNTS during the 1960s indicate the activity approached the shipments of seventeen thousand units per year reached in the late 1950s.[47]

There were, nevertheless, significant changes in the Senegalese transfusion service, despite the continuity provided by Linhard's long tenure as director. For example, the shipments of freeze-dried plasma declined dramatically in the 1960s. The reasons were many. The facility was difficult to maintain, and the advantage it offered of long-term storage was now less important because the Dakar facility could send whole blood to hospitals within Senegal. There were also reports of problems with contamination that prompted some hospitals to refuse delivery of plasma in favor of whole blood. As a result, there were few if any reports of plasma delivery by the 1970s.[48]

Another change, albeit later, was a decline in the payment for blood donation. After independence, the CNTS continued the practice established during colonial rule of paying donors 500 francs for giving blood at the center. By the 1970s, however, those whose blood was collected by the mobile van were not remunerated.[49] The rationale offered by the service was that the payment was for the inconvenience of coming to the center. Nonetheless, despite ending remuneration, blood donations from mobile donors grew significantly in the 1960s.[50]

Mobile donations remained about equal to those at the CNTS during the 1970s, and in the 1980s payment ceased altogether, as Senegal came into conformity with transfusion practices in the rest of the world. The CNTS, however, continued to offer a meal to donors. Of note also, after 1973, was a drop in collections at the CNTS, possibly the result of the economic crisis and a cut in health budgets. From 1972 to 1978 the health ministry budget in Senegal was cut by one-third (from 9 percent to 6 percent of the national budget).[51]

TABLE 3.8. Donors at CNTS, Senegal, 1960–80

Year	Total donors	Donors at CNTS	Mobile donors
1960	10,311	—	—
1961	10,114	8,609	1,505
1962	10,900	5,736	5,164
1963	8,853	5,757	3,096
1964	11,758	7,640	4,118
1965	13,490	7,414	5,076
1966	14,211	6,509	7,702
1967	9,945	9,800	145
1968	14,662	10,278	4,384
1969	11,748	5,380	6,368
1970	11,695	7,657	4,038
1971	9,795	7,085	2,710
1972	9,573	6,453	3,120
1973	6,312	3,108	3,204
1974	7,324	2,635	4,689
1975	6,444	3,180	3,264
1976	8,366	3,892	4,474
1977	6,648	3,114	3,534
1978	6,811	3,420	3,391
1979	7,366	4,869	2,497
1980	8,977	5,308	3,669

Sources: République du Senegal Ministère de la santé publique, *Statistiques sanitaires, 1963–69;* Akué, B. A. [no title] medical thesis, Université de Dakar, [1981?], 39.

Figures for the regional and district hospitals in Senegal are not complete enough to make any judgments about the long-term overall collection of blood or practices of transfusion in the whole country. Selected recent years with the most complete data reflect a significant rise in blood use, likely as a result of increased foreign assistance after the AIDS crisis.[52]

TABLE 3.9. Blood donations, Senegal, selected years, 1994–2003 (units)

Year	CNTS (Dakar)*	Hôpital Principal (Dakar)	11 regional hospitals (Senegal)
1994	8,261	6,455	8,400
2002	10,241	—	9,806
2004	12,240	5,128	11,477

Includes le Dantec.

Sources: Drs. Kabou, Boyeldieu, "Rapport de Mission de Supervision des Postes de Transfusion Sanguine," February–March 1995; Bernard Poste, "Rapport de Mission de Supervision des Banques de Sang du Senegal," March 2003; Poste, "Rapport de Mission de Supervision des Banques de Sang du Senegal," June 2004, all in CNTS Dakar.

In 1994, for example, the CNTS in Dakar collected roughly the same amount of blood as in the 1970s, and that figure was equaled in the regional hospitals (8,400 donations). Moreover, there was only moderate growth from 2002 through 2004. So, it is likely that Senegal followed roughly a similar pattern as Uganda and Kenya as far as the importance of regional hospitals outside the capital.

Taking the population of Senegal during the 1960s to be on average 3.6 million,[53] the average number of donations per year for 1961–69 was 11,459, or a rate of 318 per hundred thousand. This was lower, but still on a scale similar to Kenya (707 in 1981) and Uganda (443 in the early 1970s). Adding Senegal to an earlier chart shows that per capita, Senegal's overall donation rate by 2003 grew closer to that of the other two countries, whose donations were less able to keep up with a rapidly rising population.

Without going into as much detail, many of the features of the practice of blood transfusion after independence in Kenya, Uganda, and Senegal can be seen in other sub-Saharan African countries. Overall there was a growth in transfusion use that followed a continued expansion of hospitals both in the capitals and in outlying regions, with faster growth outside the capitals because they were starting from a low or nonexistent level of practice. Some countries, such as the Ivory Coast, Southern Rhodesia (later, Zimbabwe), and Zambia, followed a pattern similar to Kenya, Uganda, and Senegal, with national services that functioned largely in the capital and in other regions of the country as best they could. But local hospitals supplemented or complemented blood collection as use required.[54]

Other countries, especially those with scarcer resources, did not develop national services until much later, if at all. Such countries include Nigeria and Zaire, where lack of a national transfusion service was the result of logistical

TABLE 3.10. **Blood transfusion rates, Kenya, Uganda, Senegal, selected years, 1960–2003**

Country	*Year*	*Transfusions*	*Population (millions)*	*Transfusions/100,000*
Kenya	1981	113,189	16 m (1980)	707
	2001	101,212	30 m (2002)	316
Uganda	1972–74	35,412 (avg.)	11 m (1965)	443
	1995	38,854	19 m	177
	2003	102,703	25 m	395
Senegal	1960–69	11,459 (avg.)	3.6 m	318
	2003	28,805	10 m	288
U.S.	2005	15 million	287 m	5,230

Source: www.populstat.info accessed March 25, 2013. Sources of blood donations are from tables in chap. 3.

and technical difficulties, not lack of resources. Because of the size of these two countries, the hospitals developed their own methods of procuring blood that sometimes allowed them to function as regional transfusion centers, but with little coordination or direction from the national level. This combination of multiple and uncoordinated services makes it difficult to find consistent and systematic statistics about blood collection or transfusion.[55] Beginning in the mid-1970s, however, a combination of events brought great changes in national transfusion services and, accordingly, better records for analysis.

The Second Phase of Independence: Stagnation, Crisis, and Modernization after the 1970s

The 1970s marked a turning point in the history of blood transfusion in sub-Saharan Africa. This change was brought about not by a single new development, such as the appearance of HIV/AIDS in the 1980s (whose impact on blood transfusion will be examined in chapter 6). Instead, the 1970s saw a number of broader developments that reshaped the practice of blood transfusion. In fact, examining these developments helps understand the impact of something as dramatic as the AIDS epidemic in Africa.

The changes that began in the 1970s fall into two categories. First were the economic and political changes, beginning with the world recession that followed the 1973 oil crisis. Along with local civil wars and unrest in Africa these developments temporarily limited or ended the growth of blood transfusion. Second was the emergence of a model for modernization of blood transfusion services in Africa that eventually shifted it in many countries from primarily a hospital-based system to one with centralized or regional transfusion centers. Playing a key role in this new development was increased assistance to Africa from a number of more developed countries. They were joined by some well-established international organizations that adopted new policies to supply expertise and help coordinate assistance. The most important of these for support of blood transfusion were the League of Red Cross Societies (LRCS) and the World Health Organization (WHO).

Transfusion services had always competed with other health services for funds in newly independent African countries. Generally these expanded together during the 1960s, when resources were available from growing economies plus assistance from countries of the developed world. When government budgets flattened or decreased, transfusion services were unable to maintain the level of service they had achieved, let alone continue to expand. The recession that followed the oil crisis of 1973 hit most African countries hard, even though they were not large consumers of oil. The economic slowdown decreased opportunities for foreign aid from countries more immediately affected and eventually hurt African commodity sales as the world's markets shrank.[56]

African government debt increased and by the 1980s, in order to obtain new foreign loans, governments were forced by the International Monetary Fund and World Bank to accept a series of terms that came to be known as structural adjustment programs. They usually included budget cuts, an end to government subsidies, and the opening of the country's economy to the global market. Proponents argued this would balance budgets, make African economies more adaptable and efficient, and produce economic development. Critics warned of hardships and in the short run that included less access to health care.[57] In Kenya, where the 1980s were sometimes called the lost decade, structural adjustment brought the imposition of user fees and spelled the end of the 1960s' promise of medical care without charges.[58] In Senegal and Mali in the 1970s and 1980s, the expenditures on health as a percentage of the countries' overall budgets decreased to about half their former percentages.[59] Raw data from Senegal shows that the real problem was inflation. Health budgets remained the same or grew more slowly that the overall national budgets.[60]

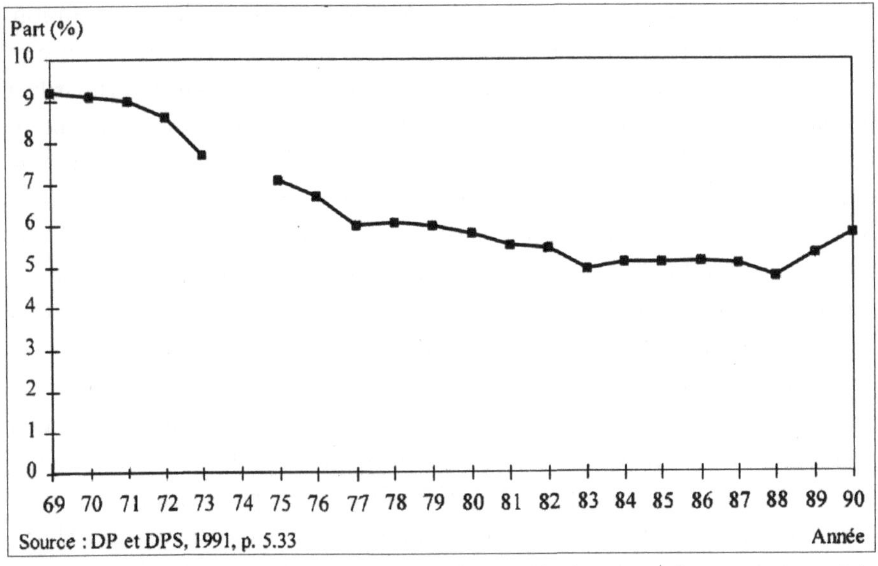

FIGURE 3.5. Health expenditures as part of overall national budget, Senegal, 1970–90. *Source:* République du Sénégal, *Tableau de bord annuel,* as cited in Brunet-Jailly, "Santé dans quelques pays."

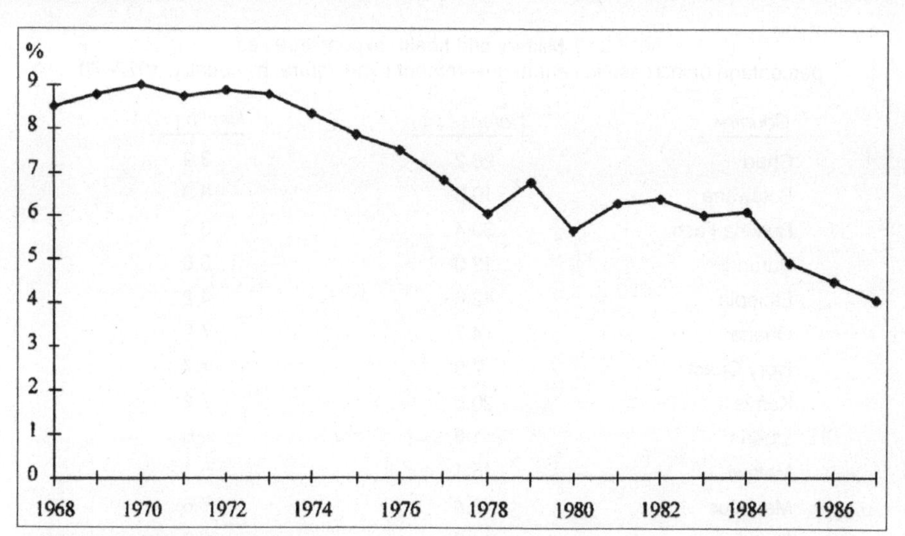

FIGURE 3.6. Health expenditures as part of overall national budget, Mali, 1968–87. *Source:* Coulibaly and Keita, *Comptes nationaux,* 147, as cited in Brunet-Jailly, "Santé dans quelques pays."

TABLE 3.11. **Health Ministry budget as percentage of national budget, Senegal, 1967–78**

Year	*National budget (billion CFA)*	*Health budget*	*Percentage*
1967–68	36.065	3.277	9.0
1968–69	36.750	3.352	9.1
1969–70	37.850	3.491	9.2
1970–71	39.000	3.556	9.1
1971–72	41.440	3.727	9.0
1972–73	44.000	3.794	8.6
1973–74	47.000	3.657	7.8
1974–75	55.000	4.103	7.5
1975–76	71.000	5.067	7.1
1976–77	86.000	5.247	6.1
1977–78	89.000	5.370	6.0

Source: République du Sénégal, Ministère de la santé publique, *Statistiques sanitaires et démographiques du Senegal, année 1977,* 8.

During the 1970s in almost every African country, similar evidence can be found for leveling or cutting health expenditures. In Uganda the recurrent government expenditures on services by the Ministry of Health rose from 52.5 million shillings in 1959 to 119.9 shillings in 1973–74, but, adjusting for inflation, the 1973–74 figure was only 34.3 million in 1960 shillings.[61] The problem was compounded by increased military spending associated with an increase in warfare and political instability in Africa during the 1970s. Africa was hardly

TABLE 3.12. **Military and health expenditures as percentage of increase in central government expenditure, by country, 1972–78**

Country	*Defense (%)*	*Health (%)*
Chad	28.2	3.9
Botswana	10.3	6.0
Burkina Faso	30.4	3.3
Burundi	12.0	3.6
Ethiopia	42.6	4.2
Ghana	4.7	7.5
Ivory Coast	7.5	7.7
Kenya	20.5	7.2
Liberia	1.6	4.6
Malawi	15.1	5.1
Mauritius	0.4	7.6
Nigeria	14.6	2.0
Sierra Leone	14.9	11.4
Somalia	17.9	5.3
Sudan	9.5	0.2
Tanzania	10.2	5.7
Zambia	—	8.3

Source: World Bank, *Accelerated Development in Sub-Saharan Africa: An Agenda for Action* (Washington, DC: World Bank, 1983), 185.

a model of tranquility in the 1960s, but fighting in the newly independent Congo was seen as largely the result of lack of planning for independence, and the Nigerian civil war was hoped to be an exceptional case. The 1970s brought more military coups than fighting in the new states, but the end result was the same: increased expenditures on the military. Data from the 1981 World Bank report by Elliot Berg that was the inspiration for structural adjustment programs showed health and military expenditure changes in the 1970s for several African countries.[62] Health expenditure rarely was over 7 percent of the national budget, whereas the defense percentage for most countries was usually over twice that.

The exact nature of budget problems varied from country to country, but even in Kenya, which like Senegal was spared civil unrest and warfare, budget problems were felt in the transfusion service, according to annual reports of declining figures for blood donation. These declines were explained by the lack of basic resources and supplies, from fuel for transportation to bags to store blood. Hence, even before the AIDS crisis of the 1980s, many African transfusion services reported declines in blood collection and transfusion.

Yet despite these problems, transfusion services in some African countries were able to modernize and expand their programs. One of the main reasons was an increase (albeit more focused) in foreign assistance, which, as is explained below, came in a wide variety of forms and amounts, ranging from the gift of motorcycles for transporting donors, to a national transfusion service with central laboratories and regional centers. Foreign assistance had been available to the new countries from the beginning. The expansion of the transfusion center in Dakar mentioned earlier was done with funds from foreign assistance only a few years after Senegalese independence. Likewise, a new Pasteur Institute opened in Cameroon the year it became independent. After the 1970s, however, international organizations of the UN and Red Cross played an increasingly important role in the development of blood transfusion in sub-Saharan Africa.

The Role of Intergovernmental Organizations in Supporting Blood Transfusion

The World Health Organization and the League of Red Cross Societies were the two most important international organizations that supported blood transfusion in newly independent sub-Saharan African countries. They did so not by providing funding themselves but by coordinating and matching potential donors with countries requesting modernization of transfusion services. They also developed standards and guidelines for transfusion, identified consulting experts, and held workshops for the purpose of sharing information. Colonial Red Cross societies had been crucial in the establishment and expansion of blood transfusion in British and Belgian colonies, but their importance diminished in many African countries after independence, as interest lapsed or governments took responsibility for the services. But in a number of smaller European countries such as Switzerland, Belgium, and Finland, plus Canada and Australia, the Red Cross ran the national transfusion services. This gave them a strong base of support and technical expertise for poorer countries, most of whom took advantage of the international coordinating organization, the LRCS in Geneva.

In 1947 the LRCS hired Zarco S. Hantchef, who eventually rose to the position of medical director in 1962. Hantchef's parents, both doctors, had emigrated from Bulgaria to France after the First World War. He received a medical degree from the University of Nancy in 1935, worked in a clinic there until the war, and after his war service moved to Geneva to work for the LRCS.

Hantchef served in various health offices of the league, where he was attracted to the work of the blood programs of many national Red Cross societies, which expanded rapidly after the Second World War. In the process he helped build, along with the International Society for Blood Transfusion (ISBT), a worldwide network within which the national leaders of transfusion cooperated.[63]

In the late 1940s and 1950s, this work consisted mainly of identifying activities and collecting information about new technology and transfusion services where they were most widely practiced, that is, in Europe, North America, other parts of the developed world, and a few newly independent countries of Asia and the Middle East. An organizational framework for this work began in 1958, when Hantchef convened the First Red Cross Seminar on Blood Transfusion, which met the week before the seventh congress of the International Society of Blood Transfusion, in Rome.[64] The society's congresses were held in alternate years, so after the second seminar, in Tokyo in 1960, Hantchef created the International Group of Red Cross Blood Transfusion Experts, which began meeting on its own in 1961 in Vienna.[65]

During the 1960s the attention of the LRCS turned more and more to the increasing number of independent countries in the world. In 1961 the Red Cross board of governors passed Resolution 8 to provide "technical and financial assistance to national societies"; in 1962 the LRCS executive committee took this idea further and passed Resolution 10, authorizing the secretary general to start development-program projects the following year in the newly independent countries. This gave Hantchef some staff to support his office to work in the area of health.[66] Support for blood donation and transfusion services was a natural activity for building national Red Cross societies in the new African countries. Some, like Uganda and Kenya, had already demonstrated these possibilities for twenty years, but most countries had not developed extensive programs, or else they had lapsed.

Hantchef developed a strategy of encouraging the local African Red Cross society in a country to work with its ministry of health to approach the LRCS to request assistance in developing transfusion services. To advise on this work, Hantchef used the International Group of Red Cross Blood Transfusion Experts, which quickly focused on two specific goals. One was to respond to the threat of commercialization of blood products, eventually culminating in the 1975 resolution "Utilization and Supply of Human Blood and Blood Products" approved by the Twenty-eighth World Health Assembly, calling for the voluntary donation of blood. The other was to provide assistance to developing countries. For the latter the experts became crucial advisers for matching requests from African countries with Red Cross societies and governments in developed countries to provide both technical and financial support. Since many of the Red Cross societies in developed countries also ran their country's blood transfusion service (e.g., Belgium, Canada, the Netherlands, Switzerland), a small group of experts, such as Alfred Hässig, director of the central service laboratory of the Swiss Red Cross blood service, served on interlocking committees that implemented Hantchef's strategy. After 1972 the transfusion experts met annually into the 1980s.

This work kept Hantchef busy as he traveled the world trying to match countries in need with those who could assist, sending expert observers to advise and assist with surveys, while also responding to even the most modest requests. For example, in 1966 the transfusion service of the main hospital in Ouagadougou, Upper Volta (now Burkina Faso), requested assistance in blood collection. Hantchef tried first to interest the Swiss Red Cross, but eventually a combination of the French government, the Canadian Red Cross, and the International Federation of Blood Donors provided blood bags (Canadian Red Cross), a motorcycle to bring donors to the hospital, and a scholarship to Paul-Henri Campaoré, a local Red Cross worker, to train for a year in France. Hantchef himself visited there in 1970.[67] Later the Swiss Red Cross provided more substantial assistance. In 1970 the Red Cross of Dahomey (now Benin) requested assistance for the blood bank at Cotonou Hospital. Hantchef referred this to the Dutch Red Cross, but no agreement was reached.[68] The following year Togo made a request to organize a transfusion center. It was referred to various countries, and $3,000 worth of equipment was provided by the Canadian Red Cross.[69] Eventually the German Federal Republic provided funding for a new transfusion center.

Not all these negotiations were successful. For example, from 1971 to 1973, Hantchef advised the Red Cross of Gambia on an ambitious proposal (£50,000) to establish a transfusion center for the country. The Swedish Red Cross responded first, sending an investigator and providing some funds for training. Eventually the Swedes found the cost too high, plus their expertise was limited because, unlike other sister societies, the Red Cross was not in charge of blood collection in Sweden.[70] Hantchef visited Zaire in 1970 as a consultant for WHO and received a number of inquiries for assistance in subsequent years. With no official government or Red Cross organization involvement, however, he was limited in responding to these requests.[71]

In the early 1970s, after the working group of transfusion experts began meeting annually to discuss the progress of the LRCS development program,

TABLE 3.13. "Contributions to the Development Programs and Donations"

Country	*Assistance*	*Amount ($ in 1976)*
Belgium	For Rwanda: blood transfusion equipment and financing of administrator	$72,228
Canada	For blood transfusion services of 13 national societies [e.g.], Togo, Tanzania, Gambia, Burundi, Rwanda	For all countries $75,000
Germany (Fed. Rep.)	For Togo: building of Red Cross blood transfusion center and provision of equipment	$100,000
Liechtenstein	For Rwanda: for payment of local staff	$16,000
Switzerland	For Upper Volta: building blood transfusion centre and provision of equipment	$100,000
	For Rwanda: 3-year extension, national blood transfusion program	$80,000
	For Burundi: extension of national blood transfusion program, also mini-bus	$26,000

Source: "Contributions to the Development Programs and Donations," Red Cross Blood Transfusion Expert Working Group, 5th meeting, Bonn, June 28–30, 1976, 5–7, AO911, IFRC. Dollar values based on historical exchange rate, June 1976, Currate, http://currate.com/historical-exchange-rates.php.

there were substantial assistance proposals.[72] According to a 1976 progress report, Hantchef was successful in obtaining assistance to develop blood programs in numerous African countries from Red Cross societies in Canada, West Germany, and Liechtenstein, in addition to Belgium and Switzerland, as noted earlier.[73]

Models of Assistance

Given all the parties involved, this process was complex and multilateral, varying greatly between countries. From the perspective of modern blood transfusion services in Africa before the 1970s, two main organizational models can be distinguished. One was primarily hospital based and developed along with the construction of large new hospitals at the end of colonial rule, as at University College Hospital in Ibadan and the new Mulago Hospital in Kampala. These hospitals were meant to be state-of-the-art facilities, including blood transfusion services. In fact, there was not always a great deal of advanced planning or thought about the support services needed for these big new hospitals, let alone that they might serve as regional or national blood transfusion centers. For example, Alan Fleming remembered that University College Hospital when he arrived, in 1962, was "one of the finest buildings of its type in the world," with well-equipped operating theaters, wards, and laboratories. "The only exception," he recalled some twenty years later, "was the Blood Transfusion Unit—it had been overlooked that the hospital would have to recruit its own blood donors, eventually to the rate of 12,000 per annum; the unit was tucked into what was meant to be part of the catering services, where it remains to this day."[74] This is a good example of how closely transfusion was tied to the growth of hospitals, and how on occasion the growth of new facilities and the doctors who staffed them attracted new patients, who drove the demand for expanded recruitment of blood donors.

The transfusion services of these hospitals provided models of hospital transfusion services that collected, tested, and stored blood using the latest techniques that could be supported with the resources at hand. Examples in capital cities include the main hospitals in Ouagadougou and Kinshasa, which secured the resources and personnel to support blood transfusion during the hospital construction boom that started in the 1960s. Foreign assistance, such as many of the requests received by Hantchef, was an important source of support, especially since helping one hospital was less expensive than creating an entire national transfusion service. In addition, as things turned out, many of these large urban hospital transfusion services soon became de facto regional blood centers, providing blood for many other, smaller hospitals in the cities and regions where they were located.

Another model that was less commonly adopted at first because of the expense was a blood transfusion service separate from a hospital with the goal of eventually serving the whole country. It might be part of a more general medical research institute or stand by itself. Among the early places in Africa where this was successful were the countries where the Pasteur Institute created branches. One in Yaoundé, Cameroon, in 1960, for example, was planned during colonial rule, and another in Bangui, Central African Republic, in 1966, was part of French assistance to a newly independent African country. Each institute had a serology section that did blood testing and functioned as a blood bank for hospitals in the capital. In Cameroon the Pasteur Institute was responsible for the rapid growth in the use of transfusion during the 1960s.[75]

The Uganda Blood Transfusion Service had been created during colonial rule as a separate facility serving a number of hospitals in Kampala. Another variation on this model that was more widely emulated in the 1970s was the dedicated transfusion center built in Dakar in 1951. As mentioned earlier, the CNTS (Centre National de Transfusion Sanguine) was planned to serve all of French West Africa, but after independence it became a national transfusion

TABLE 3.14. **Examples of international assistance for development of blood transfusion services in Africa, 1964–2003**

Year	*Country*	*Assistance from*	*Notes*
1964	Senegal	European Economic Community	Support for building expansion
1966–71	Ethiopia	Canada	Israeli adviser and technician*
1972–81	Burundi	Switzerland	Building and technician
1973	Togo	Canada	Swiss administrative technician*
1974–94	Rwanda	Belgium, Switzerland	Established a national service
1975–present	Gambia	Sweden	unsuccessful
1976–81	Upper Volta	Switzerland, Luxembourg	Support for new building
1978–84	Angola	Switzerland	—
1986–present	Mozambique	Switzerland	—
1989–99	Uganda	European Economic Community	Reestablished national service
1998–present	Kenya	U.S.	2000–2001 (after embassy bombing); 2000, Japan; later, PEPFAR
2003	12 African countries	U.S. PEPFAR	Multimillion-dollar funding for safe blood supply as part of campaign against AIDS

*In both cases the Canadian Red Cross provided funding, and the local country hired technicians from Israel and Switzerland as indicated.

Sources: See chap. 3, notes 2, 41, 67–69, 76, 79, 85, 89ff, 101–2.

service for Senegal that served most hospitals in the capital. The CNTS also served regional centers in the rest of the country as much as needs and transportation allowed. This dedicated transfusion service became a model that was emulated, at least in francophone Africa, during and after the 1970s, as foreign assistance went to institutions with a broader scope and potential yet were flexible enough to begin small and grow. The proposal by Gambia in the early 1970s followed this model but proved too ambitious. Other countries at this same time proved more successful.

Several development programs have been mentioned in this chapter; it is useful to summarize some of the other more important assistance programs for modernizing blood transfusion services in independent African countries, which became more common by the end of the 1970s. An early example was in Ethiopia beginning in the 1960s.[76] Blood transfusion began in Ethiopia as early as the 1930s, with an attempt to establish a blood bank for the hospitals in

Addis Ababa. This was cut short by the Italian war and occupation, but in the 1950s the Ministry of Health and the Ethiopian Red Cross Society (ERCS) proposed a central blood bank based in the Princess Tsehai Hospital in the capital with technical assistance from the Pasteur Institute, also in Addis Ababa. It was only in 1960, however, that the hospital organized the blood bank, under Dr. R. Ghose. He reported that by 1964 it had collected over a thousand units, but this was used almost entirely by the hospital itself.[77]

In 1966, following Hantchef's strategy, the ERCS requested assistance from the Red Cross league, which arranged for D.W. Stewart of the Canadian Red Cross Blood Transfusion Service to visit Ethiopia to investigate the country's needs. Based on Stewart's report, and after negotiations with Hantchef and the Canadian Red Cross, in 1969 Cyril Levene went to Addis Ababa. Levene, a New Zealander who headed a blood bank in Israel, immediately wrote to Stewart to see about obtaining some supplies and manuals. Before the end of the year he was joined by an Israeli technician, and together they organized the central blood bank in the buildings of the Ethiopian Red Cross Society.[78] In March 1970 the blood bank was opened by the Red Cross for the hospitals in the capital, and by 1971 over nineteen hundred units were collected and distributed. The Canadians furnished some supplies and, along with the LRCS, some published materials. The ERCS raised the majority of the funds, supplemented by a sliding-scale charge for blood units.[79]

Levene estimated that these arrangements supplied only 60 percent of the transfusion needs for the three thousand hospital beds in Addis Ababa, but the plan was that each hospital's blood collection service could fill the gap using family or replacement donors. Thus, the Ethiopian service used a central collection bank, relying primarily on voluntary donors, with hospitals using replacement donors as a base to fall back on if central supplies were inadequate. In addition to the mixed source of donors and means of collection, the plan for a national service was also a mixed system. The idea was to establish blood banks in the other big cities of Ethiopia (at the time)—such as Asmara, Harar, and Jimma—like the blood bank established in Addis Ababa, but with some flexible differences. These banks would be based in an existing hospital with laboratory facilities, but they would send personnel to the central bank in the capital for training so that standardized methods and equipment could be maintained in the country.[80]

When Levene left Ethiopia, in the fall of 1971, he reported that 75 percent of the blood collected by the central blood bank was from mobile voluntary donors. Shortly thereafter, however, things deteriorated, indicating an unfortunate feature of this period of assistance: foreign aid that was unsustainable, given the worsening budget problems of the 1970s. By March 1973 the acting secretary general of the ERCS, Dr. Asfaw Desta, forwarded an assessment he requested about the transfusion service to Hantchef. The report described in detail the decline in the service, including complaints that the organizer of the donor panel had left the Red Cross service and that the mobile collections had fallen to a minimum. The overall number of units was maintained only by an increase in replacement donors.[81] In response, the LRCS sent a number of visitors, including Hantchef himself, in January 1974. He found that the service had no director, no donor recruitment officer, and no blood centers in the provinces. Two other visitors to Addis Ababa indicated that although blood donations in the city were over three thousand units per year, they mostly came from army and police recruits or relatives of patients.[82] Thus, the success initiated by Levene was not sustained because continuing support was lacking.

After initial discussions about further assistance from the Swiss Red Cross failed, the Ethiopians approached the Finns, in part because Finland had begun supporting a program in neighboring Somalia, with whom the Ethiopians had been at war over the border region of Ogaden (1977–78). In 1981 the Finnish Red Cross sent a team whose report showed that the Addis Ababa blood bank collected approximately four thousand units, not much more than was reported in 1973. More significant was that a blood bank had been set up in Harar, which collected twelve hundred units of blood annually, but in Asmara each hospital still collected blood for its own use (also twelve hundred units total), while in other provincial hospitals between 200 and 250 units were collected.[83] By 1990, thanks to support from Finland, there were several blood banks around the country, which collected between thirty and forty thousand units of blood annually, thus supplying 90 percent of the country's needs.[84]

Assistance to the small East African country of Burundi showed two features typical of the subsequent pattern of assisting African transfusion services after the 1970s: initiation after a crisis and support for a separate national blood service, as in Senegal and Uganda. In the 1960s the Burundi blood transfusion service began on a small scale following independence. By the end of the decade, the French provided enough support to enable blood to be drawn daily in the capital of Bujumbura for use primarily by the main hospital there. In the late 1960s the Swiss Red Cross was approached by the Burundi Ministry of Health and the Burundi Red Cross for a more ambitious service, but that was cut short by the 1972 massacres and brief civil war in the country.[85] After peace was established, in July of that year, the transfusion project was quickly revived and presented to donor countries eager to assist in rebuilding. The Swiss Red Cross project, which had already been proposed, moved quickly ahead.

In late July and early August 1972, Dr. Alfred Hässig, of the Swiss Red Cross blood service (and one of Hantchef's committee of experts), visited Bujumbura and made a report back to the Swiss Red Cross about developing a national transfusion service in Burundi.[86] Shortly thereafter, on October 12, 1972, the Swiss Red Cross voted 85,000 francs credit to construct a new center. It was to be paid in three installments: 25,000 francs in November 1972; 25,000 francs in March 1973; and 35,000 in May 1973.[87] Construction began on January 19, 1973, with the completion and inauguration of the center—presided over by the minister of health, Dr. Charles Bitariho, and the president of the Burundi Red Cross, Dr. François Buyoya—taking place on December 1, 1973.

The improvements brought by the new and larger facility were quickly evident. In 1972, the last full year of the old service, there were 742 blood donations; and 1973 was similar because the new service was only completed in December (861 donations). But in 1974 there were 1,482 donations, including 367 outside the capital. In 1975 the figures were 1,667 in Bujumbura and 383 from mobile-unit collections in the countryside.[88] Impressive as these figures were, the blood was still primarily used in the capital.

At this same time in neighboring Rwanda, the Belgians entered into a much more ambitious and long-term commitment to develop a truly national blood transfusion service. Led by the Belgian Red Cross, this proved to be one of the largest, most extended, and successful of such efforts before the AIDS crisis. It was also a response to the 1972 massacres in Burundi, which caused a large number of refugees to flee to Rwanda. As in Burundi, discussions had already begun about Swiss assistance as early as 1968 for creating a transfusion center, but no action was taken. Now this project was chosen as the vehicle for a major collaboration between the Belgian development fund and the Belgian Red Cross on the one hand, and the Rwandan Red Cross and the Ministry of Health of Rwanda on the other.[89] The Swiss also contributed financial support for a while, but the Belgians had overall charge and provided by far the largest share of funding. The Belgian Red Cross Society was in charge of blood collection in Belgium, so it had the expertise to undertake the project.

Leon Stouffs, a former Red Cross employee in the Belgian Congo, who was sent to Rwanda in 1972 to help with refugees, began negotiations on the transfusion center.[90] As he had done for Burundi, Hässig, from the Swiss Red Cross, made an official assessment of conditions later in 1972, and in November 1974 Hantchef sent a proposal to the secretary general of the Belgian Red

Cross with a request for 45,000 Swiss francs for a transfusion center. By 1975 an even more ambitious agreement was signed that aimed at a transfusion service for the whole country, beginning with blood collection centers in Kigali, the capital, and Butare, home of the National University of Rwanda.[91] Three other centers were added at a later date.

The Belgian funding came initially from the Administration générale de la coopération au développement (AGCD): 2.7 million Belgian francs in 1976 (approximately 115,000 Swiss francs at 1990 exchange rates). That increased to 4.24 million Belgian francs in 1978, and was matched by 1.4 million from the Belgian Red Cross. By 1980 the two agencies annually contributed 6 million (AGCD) and 2 million (BRC), and that level of funding was maintained at least for the next five years, by which time the Rwandan government's match was 375,000 Belgian francs.[92] The results reflected the strong commitment of support. From a starting point of 286 donors at Kigali in 1976 (which does not include all the unreported local hospital donors), by the end of the first year of the new transfusion service, 1977, there were 1,922 donors. That number rose as new centers were added. One in Butare in 1980 brought the donor total to 3,240; the addition of a center in Ruhengeri brought the total to 8,756 in 1984, and in the 1986 report, 9,304 donors gave blood.[93] Note that this was even in the face of the revelation of the AIDS epidemic and the rise of donor concerns about the safety of giving blood. In one of the last full reports before the new horrors of 1994, the Rwanda transfusion service reported 22,245 donations for 1992 (384 per 100,000 population).[94] The horrors that followed shortly thereafter in Rwanda could not have been in starker contrast to this example of success in developing a blood transfusion service in an African country.

A final example of assistance to an African country from the Red Cross society of a smaller European state that also ran its national transfusion service was Finland. In the early 1980s that country began an ambitious project in Somalia and to a lesser extent ones in Ethiopia and Zimbabwe; the project in Somalia lasted over a decade. The timing coincided with the stepping down of Hantchef at the LRCS in 1980, after he reached mandatory retirement age, and his replacement by Juhani Leikola, who had previously run the Finnish Red Cross Blood Service. Just as Hantchef had relied on his close colleagues at the Swiss Red Cross, Leikola turned for assistance to someone from his home country. In this case, it came especially from Jukka Koistinen, who had been director of blood procurement at the Finnish Red Cross since 1976. He accompanied Leikola on a visit to Somalia and Ethiopia, where the Finns had earlier been involved with assistance following the droughts of 1975.[95] Koistinen developed a plan for Somalia that closely resembled the model of Hantchef, with one-third of funding from the Red Cross of Finland and two-thirds from Finnaid, the Finnish government's international development agency. Koistinen worked in Somalia over a month each year for the first three years of the project and visited annually thereafter. The financial commitment was substantial: a total of $2 million between 1982 and 1988.[96]

The World Health Organization and Blood Transfusion in Africa

The World Health Organization was increasingly consulted in these efforts of the LRCS to provide financial and technical assistance to African countries to develop their transfusion services, but WHO's role was largely technical and advisory.[97] As the services in these countries became better established, WHO, which had neither funds or services of its own, played its most important role in fostering a new form of collaboration: conferences and workshops with multiple African countries. These meetings not only provided an efficient way for outside expertise to reach a number of blood service directors but they also enabled the African personnel themselves to exchange their experiences and, more important, to control the process, as opposed to simply receiving bilateral largesse, as they had before.

TABLE 3.15. **Conferences and workshops on blood transfusion, sub-Saharan Africa, 1977–85**

Date	*Location*	*Meeting*
March 7–12, 1977	Yamoussoukro, Ivory Coast	Premier congrès africain de transfusion sanguine
March 1–8, 1978	Bujumbura, Burundi	Atelier sur l'organisation d'un service de transfusion sanguine
November 12–17, 1979	Nairobi, Kenya	WHO/LRC workshops on blood transfusion
November 7–11, 1983	Dakar, Senegal	Atelier OMS/LSCR* sur la gestion des services nationaux de transfusion sanguine
October 22–25, 1985	Bangui, Central African Republic	WHO, Workshop on AIDS in Central Africa
December 2–11, 1985	Harare, Zimbabwe	Optimal Use of Resources: African Workshop on Management of Blood Transfusion Services

*Organisation mondiale de la santé [World Health Organization]/Ligue des sociétés de la Croix-Rouge [League of Red Cross Societies]

Sources: Premier congrès africain de transfusion sanguine, 7–12 Mars 1977, Yamoussoukro, Côte d'Ivoire, actes du congrès (Paris: Institut INNIT, 1981); *Atelier OMS/LSCR sur la gestion des services nationaux de transfusion sanguine, 7–11 novembre 1983 Dakar, Senegal, Rapport* (Geneva: Ligue des sociétés de la Croix-Rouge et Croissant Rouge, 1984); *Optimal Use of Resources: African Workshop on Management of Blood Transfusion Services, Harare, Zimbabwe, December 2–11, 1985* (cosponsored by WHO/LRC/Finnida) (published reports are incomplete).

Significantly, the first of these was organized in an African country where there was little or no Red Cross presence: the Ivory Coast. Like most francophone countries, the Ivorians had followed the French model of government supervision of the transfusion service, including donor recruitment. The First African Congress of Blood Transfusion was held in 1977 in the Ivorian capital, Yamoussoukro, home of the country's president, Félix Houphouêt-Boigny, who attended the École de médecine at Dakar and worked in hospitals for a dozen years before the Second World War. Most significant, the invitation list had a blood transfusion contact in each of thirty sub-Saharan African countries. Although it is not clear how many actually attended, papers were given by representatives from six countries.[98] This was the first of several congresses, conferences, and workshops that were held almost annually in the following years. The following list indicates the participation of sub-Saharan African countries represented at conferences and workshops on blood transfusion held in sub-Saharan Africa from 1977 to 1985 (parentheses indicate if more than one conference was attended)

Benin (2)
Botswana
Burundi (3)
Cameroon (3)
Central African Republic (2)
Chad
Congo (Brazzaville) (2)
Ethiopia
Gabon
Gambia (2)
Guinea
Ivory Coast
Kenya
Lesotho
Malawi Mali (2)
Mauretania
Mozambique
Nigeria (2)
Rwanda (3)
Senegal (2)
Sierra Leone
Sudan
Swaziland
Tanzania (3)
Togo (2)
Uganda
Upper Volta (2)
Zaire
Zambia
Zimbabwe

Like their counterparts in Europe and the United States, who expanded transfusion services after the Second World War, those attending the African meetings most frequently discussed issues of a technical nature. Donor recruitment was also important, and with the recent recognition of hepatitis B as a problem, blood safety was increasingly a subject of concern. This was before the AIDS epidemic, and helps explain the ability to call rapidly organized and frequent workshops on the new epidemic as soon as it appeared. By 1985 the transfusion directors in Africa were used to holding such meetings.

Despite the complexity and differences in the histories of transfusion services in African countries after the 1970s, some general conclusions can be drawn about how they developed. First, by the 1970s there was recognition by almost every transfusion service director in Africa of the need for additional resources to make up for budget losses. This awareness was reinforced thanks to training and international conferences that informed them of new discoveries and techniques in transfusion, but all along they had been engaged in ongoing efforts to meet the obvious and fundamental needs of a national blood transfusion service. Among these were the centralized monitoring of needs, record keeping, adequate supplies, and basic screening. In other words, the goal of establishing a modern transfusion service was an ongoing concern, and requests to ministries of health for resources to do this were therefore nothing new.

The success in securing the outside funding necessary to achieve this goal usually required the national Red Cross service or a large hospital, along with the ministry of health, to contact outside funding agencies, such as WHO, the LRCS, or a development agency of a specific government. Sometimes WHO or the LRCS initiated contact with the African government as part of its campaigns beginning in the 1970s to develop transfusion services and provide technical assistance, organize workshops, and develop standards for transfusion services. The discovery of new diseases such as hepatitis B gave new impetus to this process.

If WHO or the LRCS were asked for resources, they invariably would refer the request to one or more countries or Red Cross societies with the ability and interest in providing assistance to Africa. In Europe, these were not usually the bigger countries but rather the smaller: Switzerland, Belgium, the Netherlands, and Scandinavia, plus Canada. Their foreign-aid ministries and Red Cross societies, and often their own national blood transfusion agencies, were typically the organizations involved, with the foreign-aid ministry providing the money to the national transfusion agency. Not only were these countries more interested in development, but their Red Cross societies in many cases ran their national blood programs, as with Switzerland, Finland, Belgium, and the Netherlands (until the blood scandals of the late 1990s).

Beyond humanitarianism, the motivation for assistance to Africa was sometimes linked to geostrategic goals of the cold war, maintenance of colonial ties, or, in the case of Israel, to UN votes. The specific timing of a decision to provide assistance, however, frequently followed a crisis or disaster. This can be seen as early as 1972 in Burundi and Rwanda, where new transfusion services were part of the reconstruction that followed the riots and bloodshed. Swiss Red Cross assistance to Angola and Mozambique came in the late 1970s and early 1980s, after wars of independence, and Finland assisted both sides after the Ethiopian-Somali conflict. The rebuilding of the once model transfusion service in Uganda began in 1987, after the end of military dictatorship and civil war, and in direct response to recognition of the AIDS crisis.[99] In Kenya, unfortunately, significant new resources for transfusion services came only after the bombing of the U.S. embassy, in 1998. The outpouring of offers to donate blood for transfusions of victims revealed to the Americans the inadequacy of the Kenyan National Blood Transfusion Service. In fact, collections in Nairobi in 1998 reached 11,683 units, a dramatic rise from the previous year's 4,559 units, the lowest total since the 1960s.[100] USAID agreed to supply $1.1 million in funds for modernization and centralization that had been proposed unsuccessfully for almost a dozen years.[101]

Before the AIDS crisis, outside assistance came in many forms that can usefully be divided into two categories. One was assistance for a variety of needs usually to a transfusion service or blood bank, almost always associated with a country's biggest hospital, in the capital city. Although some projects included plans for additional blood banks or centers in the rest of the country, assistance often stopped there because of lack of funding, failure to complete the initial phase on time (or at all), or more dramatic political or civil war changes. Burundi is a good example of this; likewise the Swiss left Angola before completing joint plans for modernizing the country's transfusion service.

Some projects, however, were broader from the start and were able to build regional blood centers and eventually a national service. The Belgian program that began in Rwanda in 1976 was noteworthy in continuing assistance for over fifteen years until it was stopped, unfortunately, by the disastrous recurrence of violence in 1994. Uganda in 1987, and more recently the assistance of the U.S. President's Emergency Plan for AIDS Relief (PEPFAR) to insure a safe blood supply in Kenya and several other African countries, has followed this pattern of long-term assistance to reach an entire country. These assistance programs for a national transfusion service typically began with an initial five-year project to establish the central laboratory and blood bank in order to supply blood to the capital and hospitals at some distance. Then a second multiyear plan added some regional centers, and usually a third phase was necessary to reach the whole country.

The impact of the AIDS epidemic that exploded in the middle of this process will be examined in more detail in chapter 6. As for its impact on developing transfusion services generally, at first the AIDS epidemic was yet another example of a crisis that prompted offers of assistance; but the extent of the new resources eventually became an unprecedented response to what was seen as a crisis unprecedented in scale. The assistance offered a chance to finally establish the infrastructure for what had long been the goal of African transfusion directors: an adequate and safe blood supply.

Blood transfusion services in Africa are currently in the middle of these processes of centralization and modernization, with funding being less a problem than actually developing and implementing the plans. In addition to WHO and the Red Cross (now IFRC), there is now unprecedented support from PEPFAR and other international funding and coordinating agencies.[102] Despite these new developments, the individual countries must deal with the same fundamental issues that have long been part of the practice of blood transfusion: deciding whom to transfuse, securing adequate blood supplies, and insuring safe blood. The remaining chapters will examine the history of responding to these needs.

4 WHO GOT BLOOD?

Indications for the Use of Blood Transfusion, 1945–2000

Three general conditions are necessary for transfusions to take place: patients, a supply of blood, and the knowledge and ability to do transfusions. In sub-Saharan Africa the limiting determinant was usually a lack of persons with the knowledge and facilities to do transfusions. Although early in the twentieth century there were doctors and hospitals in Africa who practiced transfusion, their numbers were far fewer than both those in need of and those willing to donate blood. Especially in the case of the former, there is little evidence of unwilling patients being an impediment to the adoption of the practice. In Africa, as elsewhere, fears and myths existed to be sure (and for the most part were allayed by practitioners), but the needs were so widespread, and the successful results of transfusions so dramatic, that if anything, the overuse of transfusion became a bigger problem than resistance on the part of patients.

Despite the obvious benefit of transfusions, which overcame reluctance on the part of Africans to donate or receive blood, the diseases and conditions for which transfusions were given in Africa have until recently received less attention than the two other features of transfusion: doctors and donors. It has generally only been since the appearance of HIV/AIDS that concern with the risk posed by unnecessary use of transfusion has prompted attention to the kinds of patients and conditions treated, and there have been almost no historical studies.[1] This chapter offers an overview based on published reports, as well as government papers and materials gathered from colonial archives, selected African blood transfusion services and hospitals, plus records of European agencies (government and Red Cross) who provided funding for the modernization of blood transfusion and screening for disease.

The Early Uses of Transfusion

Transfusions in Africa followed the practice elsewhere in the world, which meant that they were done in hospitals under the supervision of doctors. Hence, the history of transfusion follows the history of hospitals in Africa, their location and growth over time. From the start, reports of transfusions also show that they were attempted for some conditions peculiar to Africa, such as tropical diseases, especially anemia (from malaria). In fact, the first reports from the Belgian Congo, in the early 1920s, described transfusions for two conditions not typically found in Europe: an officer during the First World War suffering from blackwater fever (hemoglobinuria) and an attempt at a kind of serotherapy whereby the blood of African miners recovering from pneumonia was given to those with more severe cases.[2]

There was another and even more extensive use of transfusion in Africa before 1945 for a specific tropical condition: a systematic program to treat infants with anemia from malaria. This was first reported in 1939 in the Belgian Congo and was tried at a number of locations in the colony, most notably the hospital of the University of Louvain, in Kisantu, south of Leopoldville. There, a systematic treatment program was conducted with over fifty-seven hundred transfusions reported (an average of four per patient) from 1943 through 1951. Although it produced some success, transfusion for malaria was eventually replaced with Nivaquine treatment when it became more readily available in the 1950s.[3] The procedure continued for other problems in children, as in the late 1940s by the Lambottes at their pediatric clinic in Léopoldville, and as described in a 1965 article by Claude Bouyer at the Hôpital Sici in Pointe-Noire, Congo. He gave 2,413 subclavicle transfusions or perfusions during an eighteen-month period to infants and small children with severe dehydration caused by sickle cell anemia and hookworm infection.[4]

As medical facilities expanded in the African colonies after the Second World War, increasing numbers of doctors from Europe entered colonial health services and brought with them knowledge of simpler and more effective transfusion techniques developed during the war. The reports about transfusions in Africa focused more on these new techniques and the securing of a blood supply, with much less about the conditions of patients who were transfused. Only a few reports attempted to summarize the uses of transfusions in these early years, as in Brazzaville General Hospital, which showed a fairly even distribution of hospital services using transfusions.[5] The numbers were obviously too small to draw any conclusions, but the report did define the general categories that were most often used thereafter (see table 4.1).

TABLE 4.1. **Blood Bank transfusions, Brazzaville, 1955–56**

Medical division	*1955*	*1956*
Surgery	37	37
Obstetrics	20	29
Pediatrics	—	31
Medical	11	29

Source: "Rapport sur le fonctionnement technique de l'Institut Pasteur de Brazzaville." 1955–1956, IPO-RAP, box 11, Institut Pasteur archives, Paris.

Other striking evidence, even if indirect, of the uses of transfusion can be found in another, unusual source: the posters developed by transfusion services to recruit blood donors. These were first developed in countries with well-organized transfusion services and introduced to Africa as part of assistance offered after the Second World War. Their main purpose was naturally to help recruit donors (see chapter 5); but they also emphasized how donated blood would be used in order to persuade potential donors to give blood.

The posters usually featured large hand-drawn pictures meant to capture attention, and frequently they included depictions of those who would be the recipients of donated blood. For example, a Red Cross collection of forty-one blood donation posters from Africa between 1950 to 1980 has eighteen with prominent depictions of those who needed blood.[6] Though hardly a scientifically representative sample, the posters in that collection cover a wide range of time and geography. Therefore some recurring elements suggest they represent common uses of transfusion that were presented to Africans. Two early examples depict accident victims. One poster from Northern Rhodesia (present-day Zambia) in 1967 shows a man who has fallen from a bicycle that has crashed into a tree (fig. 4.1). The bicycle is mangled and the man has blood flowing from his head onto the ground. A smaller cartoon sequence below that image shows an ambulance picking up the victim, followed by the man in a hospital bed receiving a transfusion. The remaining panels show him leaving the hospital and others donating blood, with the message clearly spelled out: "This injured man's life was saved because someone gave him a pint of their blood. You too can help save someone's life by becoming a BLOOD DONOR." The bottom of the poster has a space to be filled in by the local blood donor recruiter, after the words "Give your name to:-."[7]

An example of an accident requiring blood transfusions for victims occurred in 1957, and it was used for a poster by the University College Hospital of the new medical school of the University of Ibadan, Nigeria. Opened in April 1957, the hospital was located only thirty miles from a serious train wreck that happened in late September 1957. The hospital's blood bank figured prominently in helping the victims, and the hospital quickly took advantage of the publicity for a mass appeal to blood donors.[8] Included was a poster that showed several train cars in a pile with bodies scattered around them, in stylistic hand-drawn form (fig. 4.2). The words "There is always a need for blood. Give yours" reflected the hospital's concern that donors needed to give blood year-round.[9]

FIGURE 4.1. Zambian transfusion poster. *Source:* BBT-1988-107-13_A003GU, MICR.

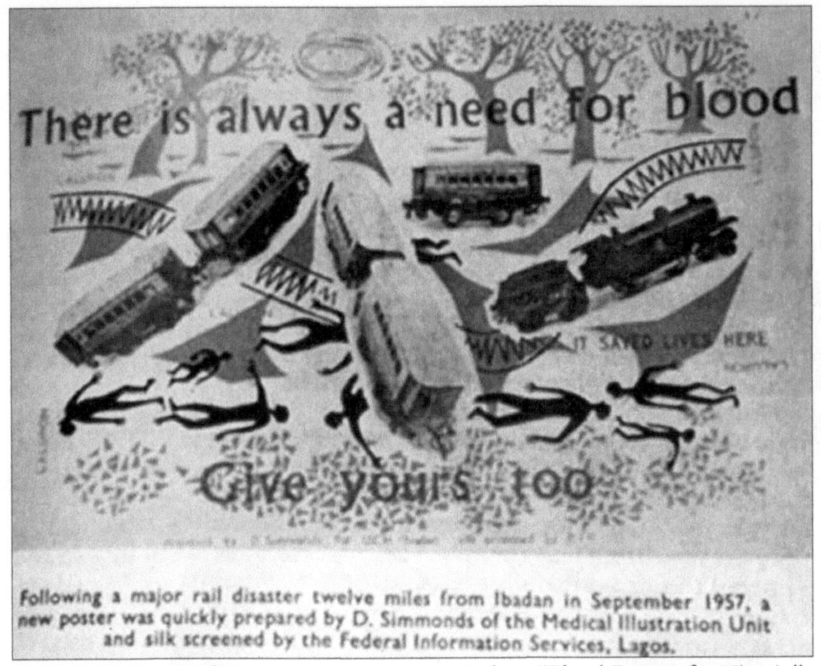

FIGURE 4.2. Ibadan transfusion poster, 1957. *Source:* Maclean, "Blood Donors for Nigeria," 29.

The need of transfusions for blood loss from accidents and other injuries was prominent in these and subsequent posters, but these needs were rarely mentioned in articles published about indications for transfusion. Hence the images in the posters are a reminder that the most common things often go unsaid. In later reports with categories indicating transfusion use, presumably this frequent condition requiring blood was counted under surgery or medicine. Subspecialties like obstetrics and pediatrics, which were newly developing in Africa in the 1960s and 1970s, received their own categories.

Posters from the late 1960s in Ethiopia were most likely used in conjunction with assistance from the League of Red Cross Societies and Levene, the Israeli advisor (see chapter 3). They also stressed the need for blood to help accident victims. One poster has a circle with four panels in the center: two panels showing donors (one a soldier), and two panels showing victims (fig. 4.3). One of the victims has been in a factory accident and is holding his bleeding arm while a coworker opens a medicine cabinet. The other setting is on a farm where the victim is already on a stretcher, with a Red Cross worker giving blood and an ambulance waiting in the background. The slogan in Amharic above and below the picture says, "Donate your blood to save a life!"[10] The sponsor is indicated at the bottom: "Red Cross Society of Ethiopia."

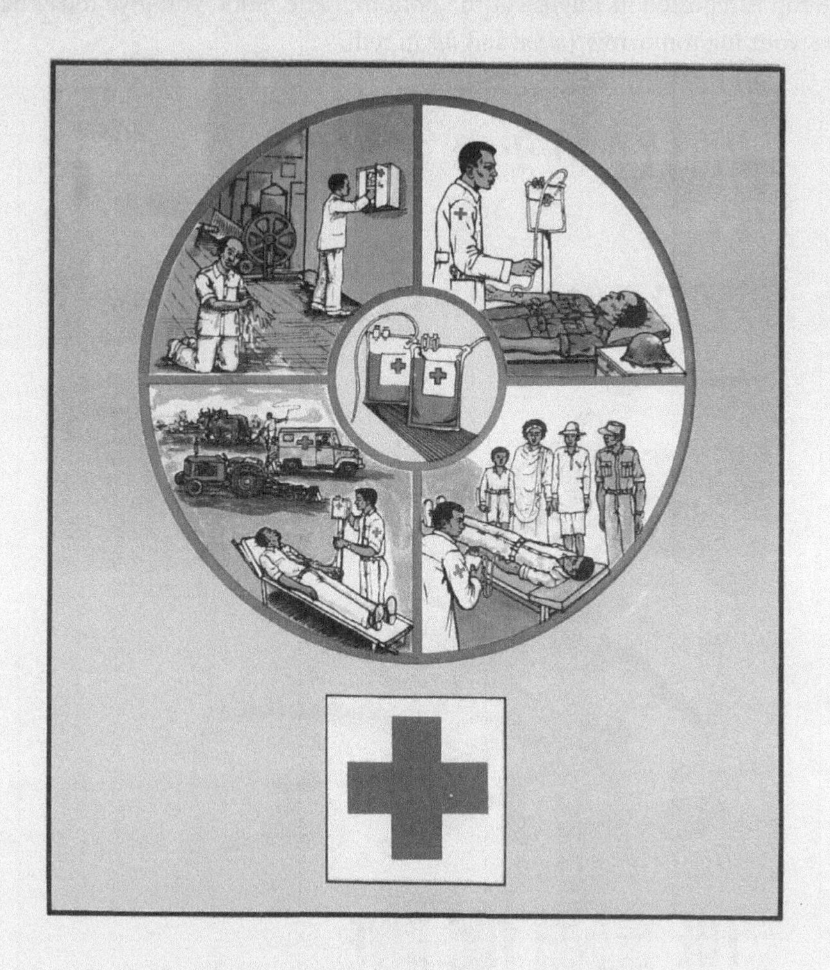

FIGURE 4.3. Ethiopian transfusion poster. *Source:* BBT-2003-35-20_a064jb, MICR.

Two other posters from Ethiopia focus on accident victims. One features a large central picture of a car that has crashed and the victim sprawled in front with his arms lying helpless at his side (fig. 4.4). A full blood bag (red) is shown at the top on the right, with a tube hanging down. The slogan in Amharic at the top is repeated in English at the bottom: "The blood you give today may save your life tomorrow [*blood* and *life* in red]."

FIGURE 4.4. Ethiopian transfusion poster. *Source:* BBT-2002–24–88_a016jb, MICR.

Another poster in this series also uses the color red liberally (fig. 4.5). At the top is a blood bag from which appear to flow four small pictures that illustrate the full range of blood transfusion uses. One shows a car crashed into a tree with the driver still inside; another shows a patient on an operating table surrounded by a doctor and nurses, with a blood bag hanging above and a tube leading down to the patient's arm. The third shows a woman, presumably

FIGURE 4.5. Ethiopian transfusion poster. *Source:* BBT-2002-16-15_a009jb, MICR.

in childbirth; and the last is a baby receiving a transfusion.[11] The slogan in Amharic at the top says, "Save a life by donating blood!!" At the bottom it concludes, "This may happen to you or your family."

A poster from Guinea gives full scope to the importance of transfusions for infants and demonstrates its possibilities for an emotional appeal for blood donations (fig. 4.6). It is a striking poster in black and white with two large images. At the top is a screaming mother with her baby in her arms, fleeing from a village hut to a small dispensary displaying a red cross. The bottom picture shows the same mother kissing a now healthy and smiling baby. The slogan in French at the bottom says, "Thanks to donated blood, my baby is saved."[12]

FIGURE 4.6. Guinean transfusion poster. *Source:* BBT-2002–27–33_a017jb, MICR.

The Liberian Red Cross also issued a series of posters using the same before-and-after format, but in this case the "before" image is a person donating blood, and the "after" shows the blood being used to save a life. One poster appealed for blood donations to save mothers in childbirth, showing in the second picture a woman, pregnant and giving birth, with a blood bag attached to her arm for a transfusion (fig. 4.7). The slogan reads, "Give blood today & save a life." Two other posters in this series repeat the accident theme. One shows a scene on a farm where one man is cutting stalks and next to him another farmer sits on the ground, holding his leg, which is bleeding, and his machete next to him (fig. 4.8). The slogan reads, "Farmers need blood. Farmers give blood. You donate blood today." The other shows a traffic accident scene, with two trucks

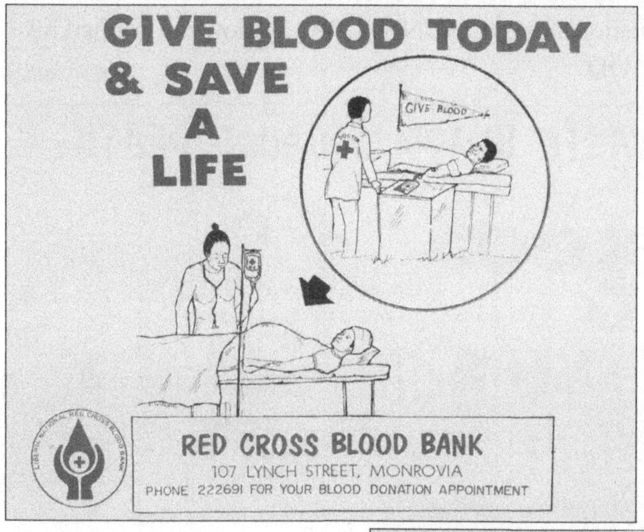

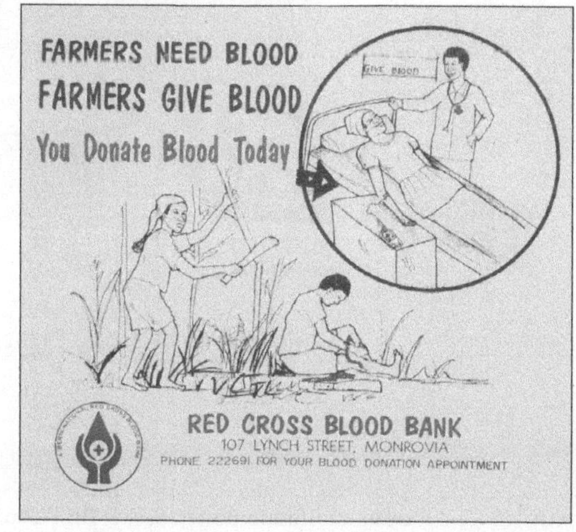

(*above*)

FIGURE 4.7. Liberian transfusion poster. *Source:* BBT-2002–29–68_a023jb, MICR.

(*right*)

FIGURE 4.8. Liberian transfusion poster. *Source:* BBT-2002–29–66_a023jb, MICR.

that have collided head-on in the background (fig. 4.9). In front, a victim lies bleeding on the ground at the side of the road, but a nurse or doctor is next to him with a Red Cross kit at his side. A Red Cross ambulance is also on the scene with two stretcher bearers coming to the victim. The slogan is quite specific: "Accidents waste blood & cause death. Give blood to save life."[13]

Another poster on the theme of the need for blood to help accident victims was produced by the Kenya Red Cross in 1973 and is a model of simplicity (fig. 4.10).[14] Seeking the maximum impact with the simplest content, the poster shows no victim. Instead, at the center of the white poster is a drawing in black of a crumpled bicycle. The front wheel is exaggeratedly large and in the form of a battered circle to attract the eye. Nothing else is in the picture—just the words above, "In the spirit of HARAMBEE [the tradition Kenyatta adopted to promote national unity], DONATE." And below the crushed bike, in red, the word BLOOD.

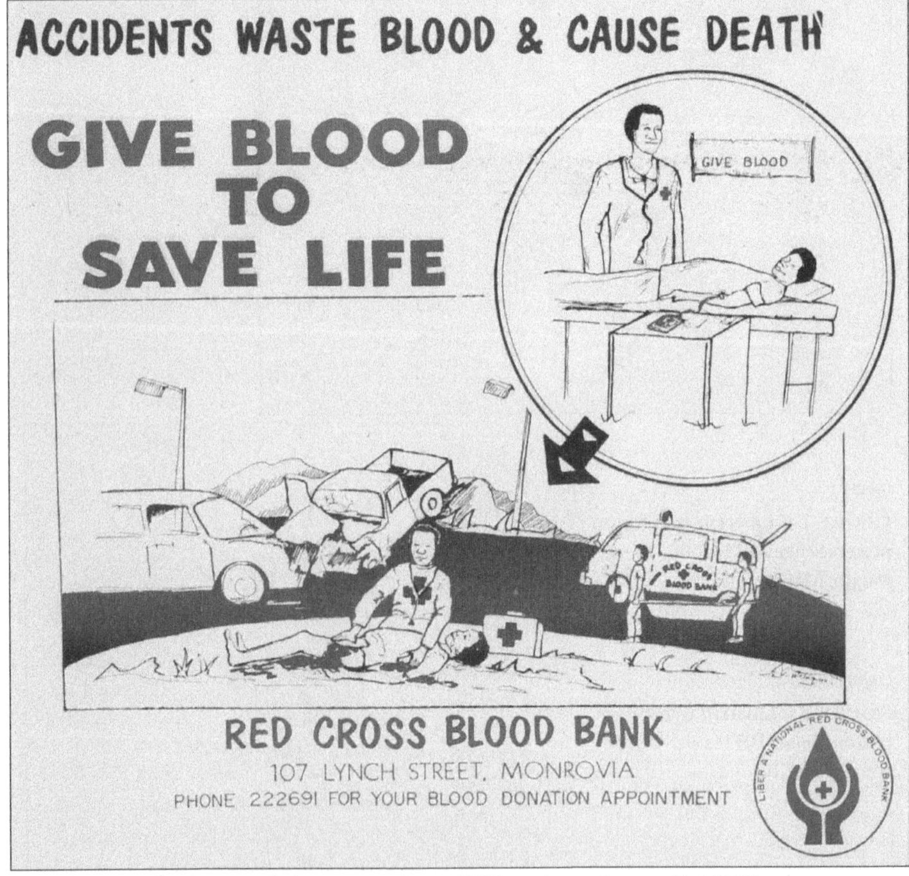

FIGURE 4.9. Liberian transfusion poster. *Source:* BBT-2002–29–69_a023jb, MICR.

FIGURE 4.10. Kenyan transfusion poster. *Source:* IFRC, AO915, IFRC.

Two posters from Angola featuring victims in need of blood also demonstrate sophistication in graphic design. One poster shows a donor at the top center, with five other pictures around him of those who need his blood for transfusions (fig. 4.11). One is of a man carrying another in his arms (though no specific injuries are shown); another victim has a bandage around an injured ankle. Two small anemic babies and what appears to be a man with war injuries are also depicted. Another poster shows more sophistication, using three images from the previous poster in striking colors (fig. 4.12).[15]

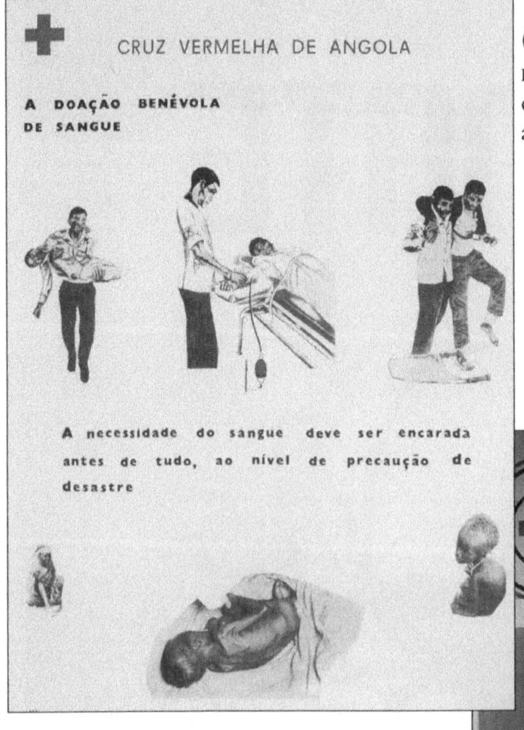

(*left*)

FIGURE 4.11. Angolan transfusion poster, ca. 1980. *Source:* BBT-2002–15–29_a007jb, MICR.

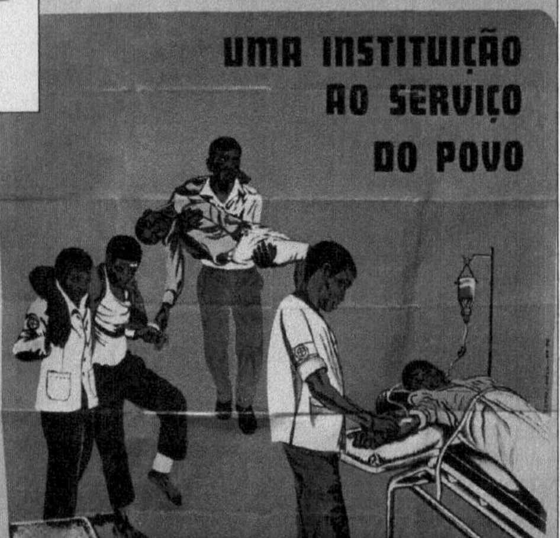

(*right*)

FIGURE 4.12. Angolan transfusion poster, ca. 1980. *Source:* 53 Angola Blutspendedienst finanzelle, korrespond., 1979–81, SRC.

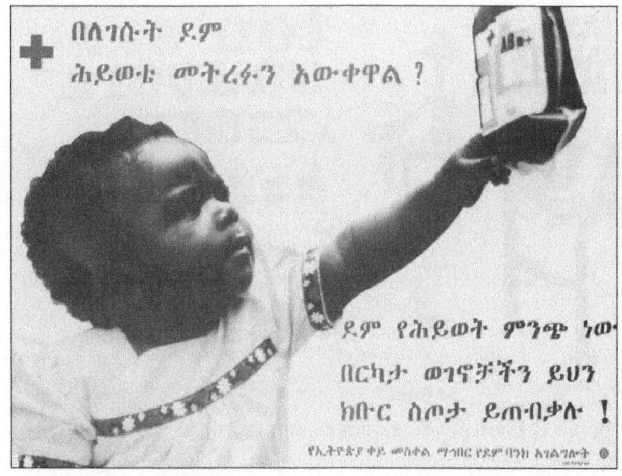

FIGURE 4.13. Ethiopian transfusion poster. *Source:* BBT-2002–16–24_a009jb, MICR.

A poster from Ethiopia also reflects several elements of technical sophistication (fig. 4.13). It features a large photograph of a young toddler, healthy in a yellow dress, who is reaching up to touch a full blood bag. There is an unusually large amount of text, in Amharic. At the top are words of the young girl, "Did you know that you saved my life by donating blood?" Below her arm, the poster reads, "Many citizens are waiting for this precious gift!" And in small print the sponsoring agency: "Blood Bank Service of the Ethiopian Red Cross Society."[16]

Finally, two other posters featuring transfusion recipients are noteworthy, one for its style and the other for its origin. The first, produced by the Togolese Red Cross, is most unusual, almost unique, as a stylized poster (fig. 4.14). It consists of a black background with striking images in white and color of two figures whose arms are locked together, with a direct red blood line connecting their hearts. The figure on the left is the healthy donor and the figure on the right has a bandaged head and left arm in a sling. Above is what appears to be a flask filled in red, with lettering around it that reads, "Donnez votre sang pour sauver [Give your blood to save]."[17]

The final two posters were produced in Benin by schoolchildren who won a contest for transfusion posters that were published in a

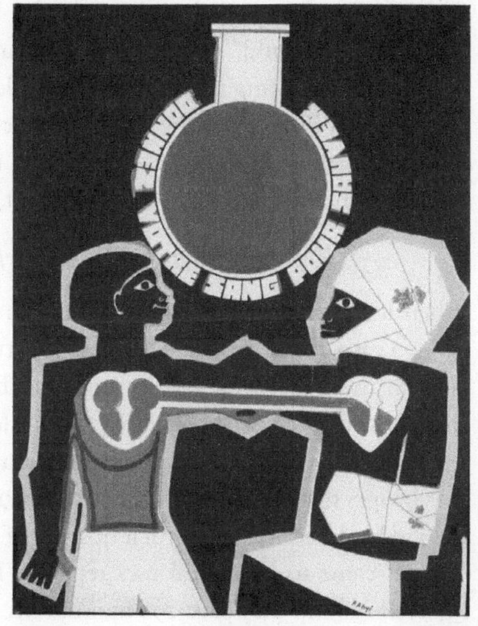

FIGURE 4.14. Togolese transfusion poster. *Source:* BBT-2002–27–62_a018jb, MICR.

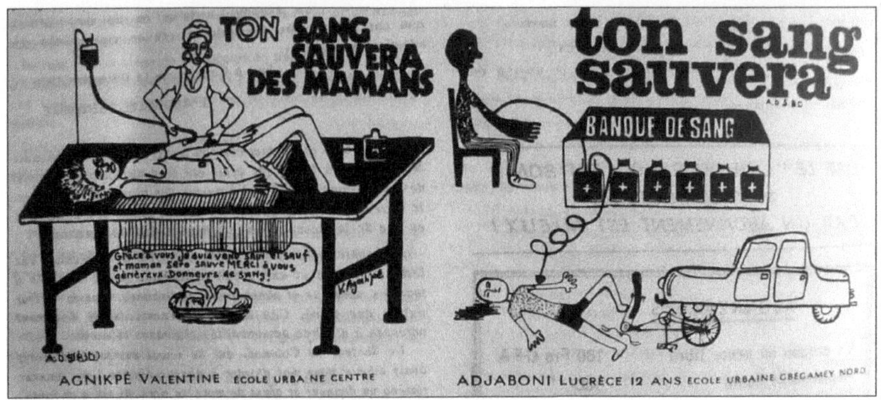

FIGURE 4.15. Benin transfusion poster contest. *Source: Le lien du sang* (1976), p. 12, AO73611 Benin, IFRC.

1976 newsletter of the Voluntary Blood Donors of the People's Republic of Benin (fig. 4.15).[18] The poster on the left shows a mother on a table in child-birth, a nurse attaching a transfusion bag to her arm. The baby in a basket under the table is saying, "Grâce à vous je suis venu sain et sauf et mama sera sauvé. Merci à vous généreux donneurs de sang. [Thanks to you I was delivered safe and sound. And mama will be saved. Thank you, generous blood donors.]" The slogan at the top reads, "Ton sang sauvera des mamans. [Your blood will save mothers.]" The poster is signed by the student; the school's name appears next to the student's printed name. The second poster depicts a donor seated in a chair, a tube connected from his arm to a bottle in a building labeled "banque de sang [blood bank]." Five other bottles are in the building, but coming out of the same bottle being filled is a tube that leads down to another child who has fallen from a bike struck by a car, blood coming from the head, and the tube of blood leading to the arm. The slogan reads, "ton sang sauver [your blood saves]"; at the bottom, next to the student's name, is her age: twelve.

Reports of Transfusion Use before the AIDS Epidemic

When transfusions became more common in Africa, after the 1960s, there were enough cases for analysis of use. These reports began intermittently then increased through the mid-1980s, that is, until the discovery of the AIDS epidemic. It is difficult to draw general conclusions, but a few representative and more extensive reports will give a better idea of transfusion use and serve as a base for understanding changes that followed. For example, a 1962 report by two doctors who set up a blood bank service in the rural hospital of Bumbuli, Tanganyika, indicated some of what came to be typical uses reported for transfusion.[19] Over the course of eighteen months they gave 294 transfusions, more than half of them for "anemia."

TABLE 4.2. Transfusions, Bumbuli Hospital, Tanganyika, 1961–62

Condition or medical division	*Cases*
Anemia (primary or secondary)	141
Pre- and postoperatively	126
Burns	4
Diarrhea	4
Undefined	3

Source: Walter and Langlo, "Blood-Bank Service," 702–7.

A study of patients receiving transfusions in an internal medicine service in Dakar, Senegal, between January 1970 and April 1973, focused on the question of transfusion accidents in 297 patients receiving one or several transfusions (approximately sixteen hundred units).[20] Although not the main focus of the article, the authors reported the following conditions of patients requiring transfusion:

TABLE 4.3. Transfusions, Senegal Medical Service, Dakar, 1970–73

Condition or medical division	*Cases*
Anemia	
from ancylostomiasis (hookworms), drepanocytosis (sickle-cell), G6PD deficiency, and chronic kidney failure	179
Hemorrhage	
from gastroduodenal ulcers, rectum, and espistaxis (nosebleed)	97
Cirrhosis	21

Source: Sankale, Ruscher, and Touré, "Accidents et incidents," 307–9.

More studies appeared in the 1970s specifically about indications for patients because they were being transfused in larger numbers and outside assistance required more detailed reports. One example was the project in Burundi supported by the Swiss Red Cross, which provided assistance to the Ministry of Health and the Burundi Red Cross beginning in 1973.[21] Within a year this project doubled the number of units collected in Bujumbura to over fifteen hundred annually. In a report of the use of blood collected for the two big hospitals in the capital and mission stations in the vicinity, obstetrics and pediatrics were the services demanding the most transfusions, although surgery and internal medicine were not far behind (see table 4.4). This report is noteworthy because it used categories that later became standard.

TABLE 4.4. Transfusions by hospital service, national blood donor service, Burundi, 1974–75

Condition or medical division	*1974 (%)*	*1975 (%)*
Surgery	231 (15.6)	398 (23.8)
Obstetrics	337 (22.7)	512 (30.7)
Pediatrics	613 (41.4)	481 (28.9)
Internal medicine	301 (20.3)	276 (16.6)
Total	1,482	1,667

Source: Schindler, *Burundi Red Cross*, 7–8.

Records of blood transfusions given during a ten-month period at Mulago Hospital in Kampala, Uganda, also show a very high use by the obstetrics service.[22]

TABLE 4.5. **Transfusions by hospital service, Mulago Hospital, Kampala, Uganda 1974–75**

Medical division	*Cases (%)*
Surgery	205 (24)
Obstetrics	409 (48)
Pediatrics	181 (21)
Internal medicine	58 (7)
Total	**853**

Source: Personal papers of Professor Esau Nzaro, Senior Consultant, Mulago Blood Bank, Kampala, May 2006.

By 1980 the numbers of transfusions were at their peak, and directors of African transfusion services began meeting at conferences and workshops, where they could share information. One Nigerian hematologist who was active in these meetings was Aba Segua (Sagoe) David-West at the hematology department of University College Hospital in Ibadan. In 1980 she presented a study of a much larger number of transfusion cases than previously reported. Her hospital had six hundred beds, and in an extensive article she analyzed the source and amount of blood requested, response of the blood bank, and eventual amount of blood used.[23]

TABLE 4.6. **Transfusions, University College Hospital, Ibadan, Nigeria, 1980**

Medical division	*Units requested*	*Units cross-matched*	*Units transfused*
Surgery	7,826	5,244	2,241
ObGyn	8,044	6,468	2,344
Pediatrics	3,215	2,908	2,183
Internal Medicine	2,011	1,290	634
Total	**21,096**	**15,910**	**7,402**

Source: David-West, "Blood Transfusion."

According to David-West, despite great demands by surgery and obstetrics, there were fewer units used because operations and deliveries proved to be safer and presented fewer complications than expected. The generally higher rate of requests, she suggested, occurred because the Ibadan hospital was a referral hospital with the highest concentration of specialist doctors in Nigeria. The request rate per hospital bed was reported at forty-two units per bed per year, and the transfusion rate was fifteen units per bed per year. David-West claimed that the corresponding transfusion rates at other hospitals in Nigeria were considerably lower.

TABLE 4.7. Transfusion rates, Nigerian hospitals, 1980

Hospital	*Transfusion rate (units/bed/yr)*
UCH, Ibadan	15
Other state specialist hospitals	5–10
District hospitals	1–8

Source: David-West, "Blood Transfusion."

Instructive as these examples are, they are limited in time and location and thus offer only snapshots of the use of blood transfusion in sub-Saharan Africa. Beginning in the 1970s and more so after the 1980s, there are better records available that follow the use of transfusion over a longer period of time and in different locations in Africa; hence, they provide an indication of trends. One was a report about blood typing before transfusion in the 1970s from Yalgado Ouédraogo Hospital, the largest hospital in Ouagadougou, Burkina Faso.[24] These records were available because in the late 1960s the hospital, the country's Red Cross society, and its Ministry of Health contacted Zarco Hantchef at the LRCS for assistance with blood transfusion in conjunction with its development program. Hantchef visited there in 1969, and in subsequent years a number of Red Cross societies, governments, and NGOs provided a wide range of assistance to the hospital, culminating in 1983 with construction of a blood bank at the hospital with funding from the Swiss Red Cross.[25] As a result the hospital was required to keep records of transfusions to report on how the assistance was helping.

Figure 4.16 does not show actual transfusions, but rather requests for blood from hospital services. Thus, the numbers indicate changes in demand

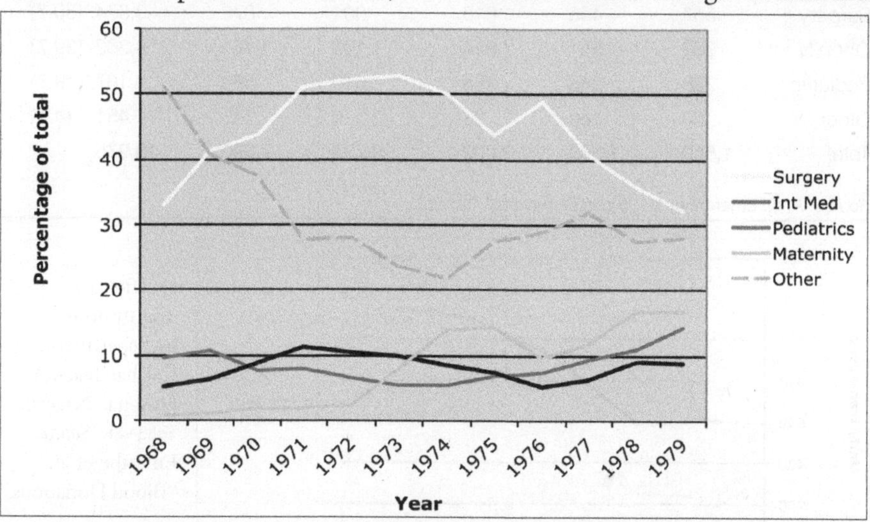

FIGURE 4.16. Transfusion requests by department, Yalgado Ouédraogo Hospital, Ouagadougou, 1968–79. Source: "Dons reçus dans le cadre du Programme de Développement, 1963/73," IFRC, AO735/2 Upper Volta, IFRC.

for blood over time. These are unusual because they show a surprisingly low percentage of requests from maternity and pediatric wards. It may be that, in the case of obstetrics, some transfusions occurred as emergencies that were handled in the operating room and not reported. Or that pediatrics was still developing as a subfield. (A pediatric wing was added in the early 1980s.) A later report of blood use in the Ouagadougou hospital during 1983 showed 39.7 percent in maternity, 27.7 percent in surgery, and 17.7 percent in pediatrics, figures more comparable to other countries.[26]

Another run of several years of data on transfusion use was reported by hematologists at the blood bank of the four-hundred-bed University of Calabar Teaching Hospital in southeastern Nigeria.[27] They reviewed the registers for blood donors and recipients from 1984 through 1988, which is of interest because that five-year period coincided with the beginning of the AIDS crisis (see table 4.8). The most surprising feature of this data is that the AIDS crisis had no apparent effect on the use of blood transfusions, which almost doubled in those five years.[28] The proportion of use by hospital service is even more clearly displayed in figure 4.17. Although the use of transfusion increased in all services, its use in childbirth grew much more rapidly. The biggest increase was in pediatrics, although it still used less than 22 percent of blood units in 1988.

TABLE 4.8. **Transfusions, by hospital division, Calabar, Nigeria, 1984–88 (units)**

	1984	*1985*	*1986*	*1987*	*1988*	*Total (%)*
Medicine	74	78	15	140	158	621 (5.7)
Surgery	587	496	618	801	872	3,374 (30.7)
ObGYN	603	581	914	1,103	1,161	4,362 (39.7)
Pediatrics	228	248	488	619	584	2,167 (19.7)
Other	68	160	38	70	115	451 (4.1)
Total	**1,560**	**1,563**	**2,209**	**2,733**	**2,890**	**10,975**

Source: A. O. Emeribe et al., "Blood Donations," 330–32.

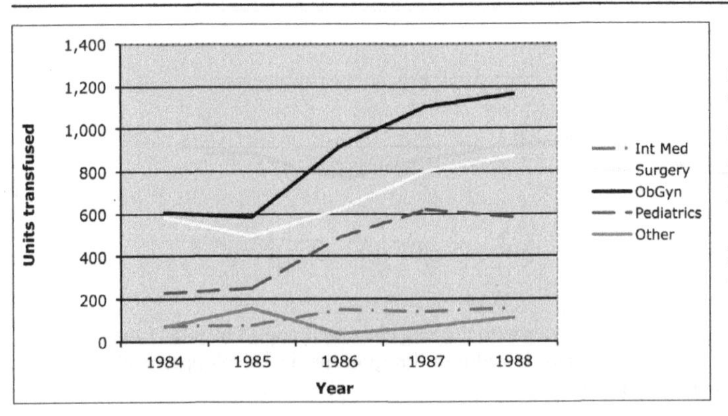

FIGURE 4.17. Transfusions by department, Calabar Teaching Hospital, Nigeria, 1984–88. *Source:* Emeribe et al., "Blood Donations."

The heavy use of transfusion for pediatric patients at Mama Yemo Hospital in Kinshasa was included in a 1988 study of 167 pediatric patients. Of these, 112 had malaria and 78 of them received transfusions. As background, the study reported that from 1982 through 1985 the total number of pediatric transfusions annually (~1,800) was typically double that of transfusions in obstetrics/gynecology, surgery, or internal medicine services. Then in 1986 pediatric transfusion spiked to over 5,000 when a chloraquine-resistant strain of *Plasmodium falciparum* appeared in Kinshasa.[29]

The longest run of information on uses of blood transfusion comes from the records of the Belgian Red Cross, which provided assistance to the government of Rwanda from 1976 to 1993 to establish a countrywide blood-banking system. Almost complete and relatively consistent categories were used that provide an excellent indication of the changes in the use of transfusion for this eighteen-year period.[30]

TABLE 4.9. **Uses of transfusion, Rwanda, 1976–93 (percentage)**

Year	*Surgery*	*ObGyn*	*Pediatrics*	*Int. med.*	*Other*
1976	28.6	35.5	10.5	4.0	21.4
1977	30.5	29.2	7.2	6.3	26.8
1978	24.1	35.1	5.0	4.3	31.5
1979	11.8	37.1	7.5	6.2	37.4
1980	—	—	—	—	—
1981 Butare	40.0	37.0	16.6	4.9	1.5
1982 Kigali	32.5	29.6	13.5	14.1	10.3
1982 Butare	35.2	43.1	14.9	6.2	0.6
1983	34.9	34.6	16.0	13.5	1.0
1984	38.5	35.7	9.6	15.4	0.8
1985	34	33	14	14	5
1986	—	—	—	—	—
1987	33.0	29.4	—	—	37.6
1988	30.5	24.2	26.2	18.7	0.4
1989	33.7	26.0	26	14	0.3
1990	30.5	22	32.5	15	—
1991	28.5	22.5	32	17	—
1992	27	24	36	13	—
1993	25	21	34	20	—

Source: "Projet Inter-Croix-Rouge de Transfusion sanguine au Rwanda, Rapport annuel," 1976–93, BRC Mechelen.

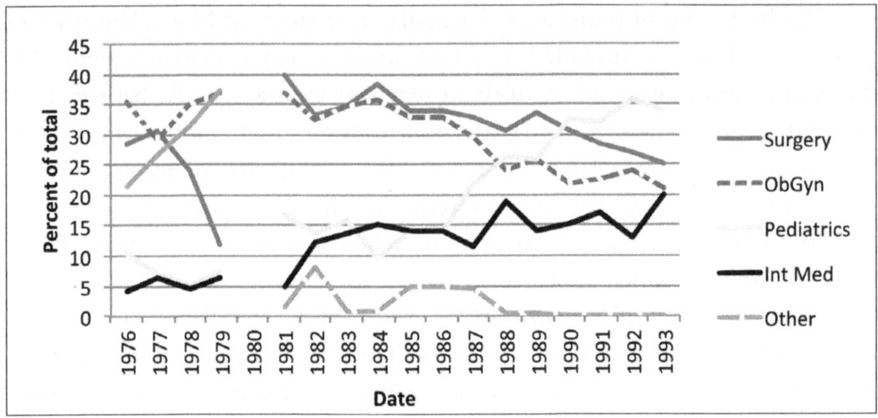

FIGURE 4.18. Transfusion use by hospital department, Rwanda, 1976–93. *Source:* "Projet Inter-Croix-Rouge de Transfusion sanguine au Rwanda, Rapport annuel," 1976–93, BRC, Mechelen.

A graph illustrates even better the rise and fall of these uses. Figure 4.18 shows that the hospital services that used transfusion most frequently at first, surgery and ob-gyn, gradually declined, while internal medicine, and especially pediatrics, increased their share of blood use. Pediatrics surpassed all others in 1990 and used 34 percent of blood units in 1993, the last full year of records.[31]

Two final examples, from Cameroon, offer a caution about drawing broader conclusions from individual hospital data. At the small, rural Protestant hospital of Mbuou (near Bafoussam, capital of West Province), around three hundred transfusions were done per year between 2002 and 2008.[32] Yet as the following chart indicates, this small hospital with limited resources used transfusions primarily for emergency cases. This hospital did not have a large enough surgery or pediatrics department to warrant other than occasional transfusions as part of those services. By far transfusions were mostly used in emergencies, in fact accounting for between 60 and 80 percent of them.

TABLE 4.10. **Blood donations, Protestant Hospital of Mbuou, Cameroon, 2002–8**

Year	*Total units*
2002	290
2003	327
2004	282
2005	367
2006	228
2007	252
2008	319

Source: Kamdem, "Évolution de la pratique."

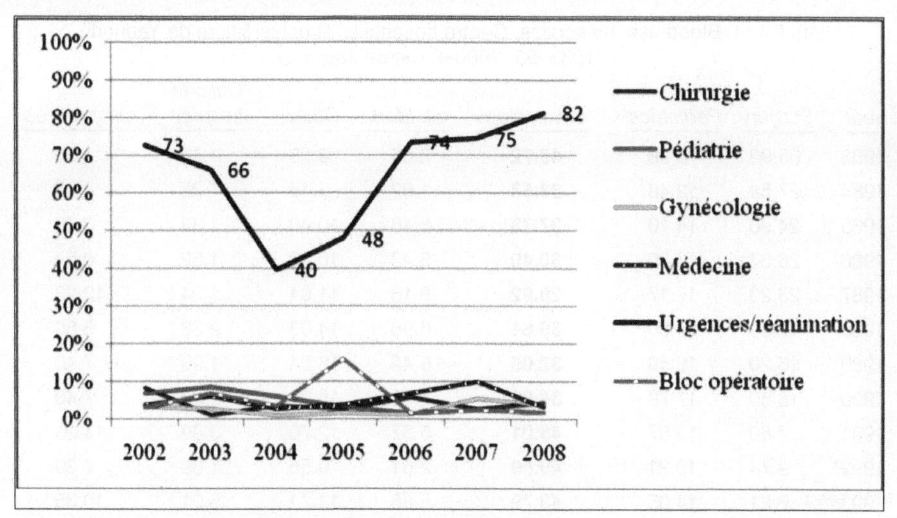

FIGURE 4.19. Transfusion use by department, Protestant Hospital of Mbuou, Cameroon, 2002–8. *Source:* Kamdem, "Évolution de la pratique."

At the large teaching hospital of the Medical School of the University of Yaoundé, far more blood was collected, since it was located in the capital and, as the country's only medical school, possessed an advanced laboratory and blood bank.[33] Because of its special responsibilities as a teaching and referral hospital, blood was used in a different way there. Sandrine Simeu Kamdem observes in this case, "the gynecology service received the most blood from 1983 to 2000. From 2003 to 2008 outside services benefited the most. It is of interest to note that use for pediatrics was not very important."[34]

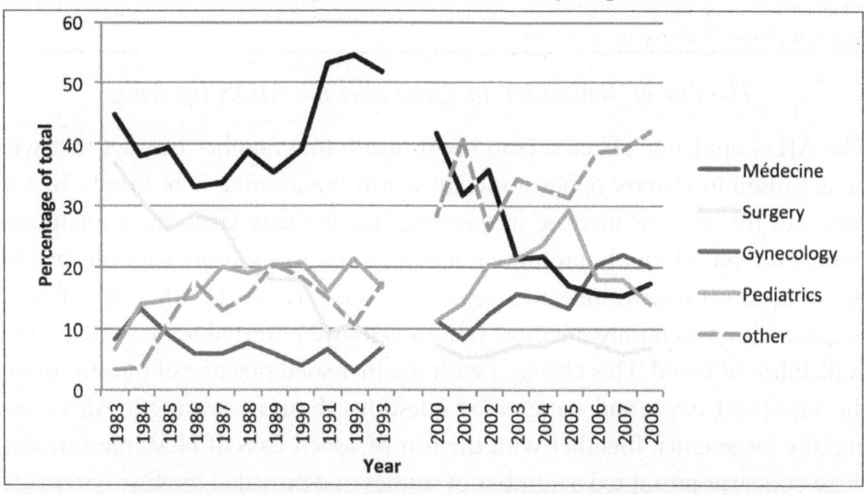

FIGURE 4.20. Blood donation, Medical School, University of Yaoundé, 1983–92, 2000–2008. *Source:* Kamdem, "Évolution de la pratique."

TABLE 4.11. **Blood use, by service, Centre hospitalier et universitaire de Yaoundé, 1983–93, 2000–8 (percentage)**

Year	*Surgery*	*Pediatrics*	*Gynecology*	*Int. Med.*	*Other*	*Outside hospital*	*Unspecified*
1983	35.93	6.28	**43.72**	8.06	3.28	1.43	1.30
1984	27.54	12.48	**33.53**	11.62	3.16	7.25	4.43
1985	24.90	14.10	**37.73**	8.46	10.40	1.32	3.09
1986	26.37	13.89	**30.40**	5.43	16.25	1.62	6.04
1987	23.23	17.17	**28.92**	5.18	11.04	1.24	13.23
1988	17.89	17.30	**35.64**	6.96	14.03	2.38	5.80
1989	16.20	18.49	**32.05**	5.42	18.54	1.90	7.40
1990	15.60	17.79	**34.02**	3.32	15.94	5.93	7.40
1991	7.80	13.67	**45.01**	5.57	12.70	3.99	11.25
1992*	9.74	19.21	**49.59**	2.61	9.56	1.89	7.39
1993	6.21	14.05	**43.79**	5.88	14.71	5.01	10.35
2000	5.81	9.10	**33.79**	9.10	23.01	12.46	6.73
2001	4.20	10.92	24.94	6.49	**31.95**	17.47	4.02
2002	3.96	14.67	**25.67**	8.87	18.78	25.42	2.63
2003	4.83	14.54	14.50	10.63	23.29	**29.39**	2.81
2004	5.15	16.66	15.35	10.50	23.12	**25.19**	4.04
2005	5.86	19.53	11.15	8.81	20.82	**31.01**	2.81
2006	4.62	11.22	9.69	12.79	24.23	**30.91**	6.55
2007	4.03	12.03	10.22	14.83	26.28	**30.35**	2.26
2008	4.25	8.32	10.35	11.96	25.30	**34.40**	5.42

Notes: Bold type indicates highest percentage use. Due to rounding, percentages do not always total 100%.

*11 months.

Source: Kamdem, "Évolution de la pratique."

The Use of Transfusion in Africa after the AIDS Epidemic

The AIDS epidemic affected blood transfusion in a number of ways that will be discussed in chapter 6, but some are worth mentioning here briefly. In addition to the obvious increase in screening for the new virus, the contamination of the blood supply prompted doctors to be much more cautious before deciding to do transfusions. In a sense, this was a return to when transfusion began, a time when only the most serious cases were treated because of limited availability of blood. This changed with the increased practice of transfusion in the 1970s and 1980s, and some raised questions about its overuse in Africa, especially for anemia. Together with the fear of added risks of HIV transmission, these concerns prompted a number of studies that provided the first systematic assessment of blood use in sub-Saharan Africa.

One of these studies examined clinical records for a three-week period during July and August 1990 in three hospitals in Kumasi, Ghana. The importance of this research was not the numbers (only 278 cases) but the attempt to determine unnecessary transfusions. The main criterion used for this judgment was transfusion of one unit or less of blood because that amount was considered too little to be effective.[35] Some additional transfusions were judged unnecessary because of inadequate tests or vague clinical findings (e.g., weakness, pallor, palpitations). The authors found that 17 percent of transfusions were unnecessary.[36] A 1991 study of two hospitals in Abidjan, Ivory Coast, was similarly concerned with overuse of transfusion. It listed three main reasons for overconsumption of blood: doctors ordering too much blood for fear of running short, overuse of transfusion in caesarian deliveries, and treatment of parasitic or nutritionally caused anemia by transfusion, rather than finding and treating the underlying cause.[37]

Studies done on transfusion use in the 1990s in Kenya (Siaya District Hospital), Mozambique (Maputo Central Hospital), and Cameroon (Central Hospital of Yaoundé) confirmed earlier patterns of use.[38] For the studies with comparable data the following results are for blood transfusion use by hospital service. The examples from Ghana, Mozambique, and Cameroon showed a similar high use of transfusion for pediatrics, while the Abidjan hospitals showed an even balance between services.

TABLE 4.12. **Blood units used for transfusion, by service, various African hospitals, 1990–98**

Hospital service	*Kumasi, Ghana, July–Aug 1990*	*Abidjan, Ivory Coast, Jan 1991*	*Maputo, Mozambique, 1992*	*Yaoundé, Cameroon, 1994–98*
Surgery	29 (10%)	321 (22%)	88 (17%)	(see Int. Med.)
Ob/Gyn	60 (22%)	306 (21%)	85 (16%)	4,855 (18%)
Pediatrics	152 (55%)	370 (26%)	246 (47%)	17,128 (63.5%)
Int. Med.	37 (13%)	231 (16%)	20 (4%)	4,990 (18.5%)*
Other	—	225 (15%)	82 (16%)	—

Source: Addo-Yobo and Lovel, "Preventing latrogenic HIV"; Mignonsin et al., "Transfusion sanguine," 723–31; Barradas, Schwalback, and Novoa, "Blood and Blood Products"; Mbanya, Binam, and Kaptué, "Transfusion Outcome."

In sub-Saharan Africa the practice of blood transfusion historically followed the spread of hospital-based health services and the increased number of trained physicians. Hence, transfusion was first used in the medical procedures most immediately needed by patients, especially surgery and obstetrics. The history of pediatric use, and in particular for anemia from malaria, leads to two additional conclusions. First it reflects the peculiar problem with malaria in Africa, for which cure and prevention have failed to meet rising needs.[39] Hence the increased use of transfusion, at least to stabilize an infant with severe malaria-related anemia reflected the lack of other treatments for this particular health problem in many parts of the continent. Second, the pediatric use of transfusion became widespread only after the procedure was established as a dramatic, life-saving measure for sudden loss of blood and imminent death. The limited facilities and personnel necessary for transfusion could not at first meet additional, larger demands. The reports of increased pediatric use of transfusion after the 1970s coincided with the rise in overall use of transfusion in most African hospitals at the district level, with regular blood supplies obtained from national or regional services or from family and friends as needed.

Since the 1990s the question of transfusion use in Africa has been much more frequently raised because of the suspicion of overuse and the greater risks associated with the AIDS epidemic. Those features of use will be examined further in the subsequent chapter, on the history of transfusion risk.

5 WHO GAVE BLOOD?

The previous chapter showed how the uses of blood transfusion were publicized in order to encourage blood donation. In Africa, as elsewhere, the question of who gives blood is of great importance for at least two reasons. First, in order to secure an adequate supply of blood it is crucial to know who gave and, even more important, why blood was given. Second, who donates blood is very relevant to the safety of blood transfusion, because people in certain circumstances have higher risk of carrying infections. So the question of who gave blood in the history of transfusion in Africa is worth examining in more detail not only because donors were one of the three crucial elements needed for transfusion but also because the answer carries lessons for current practice. First, it is useful to dispel some of the myths and stereotypes about who donated blood in Africa.

Taboos, superstitions, and cultural traditions prevented many Africans from donating blood. Like peoples and cultures around the world, Africans recognized the special importance of blood and were sensitive to its loss being connected to diminished spirit, force, or life. One revealing example was the circulation of rumors about "vampire men" in Central and East Africa beginning in the 1930s that have been compiled and analyzed by Luise White and others.[1] They show the persistence and wide spread of stories about Europeans and some Africans taking blood for drinking, making medicines, and other purposes including transfusion. White's goal was not so much to find the source or origins of the rumors as much as to reveal how limited our understanding is of African perceptions of new developments during colonial rule, including new medical procedures.

White finds that injections and drawing blood in conjunction with such things as British medical campaigns against the tsetse fly were part of the earliest stories, and the taking of blood for transfusion, especially the requirement that family members of patients give blood, also became part of the accounts she collected.[2] For purposes of this book, these studies demonstrate the importance of blood in African societies and an indication of the awareness of transfusions. They are also a reminder that this study is limited to documenting when, where, and how much blood was donated and transfused, not why it was transfused or how well Africans understood the process.

The appeal of Western medicine was strong nonetheless, and as previous chapters have shown, despite rumors and misunderstanding or the warnings of old colonials, ultimately Africans were persuaded to give up their blood and agreed to place the blood of another person in their body. From a broader perspective, Africans were generally not so different from potential donors and patients in Europe or America who had reservations and questions about giving blood and receiving transfusions. In fact, lack of blood was not the biggest hindrance to the growth of blood transfusions. Africans gave blood and welcomed blood transfusions, eventually almost to the fault of overuse.

Along with these similarities, there were some important circumstances in Africa that made blood donation different from elsewhere. They included a poor economy as well as diseases that required different treatment and specific use of transfusion. For example, a payment and meal or other refreshments in the African context meant something far more than small payments in some European countries for donating blood. Payment for blood in Africa was not a token compensation for inconvenience, as it was often described in France, for example. It was a source of cash not otherwise readily available to most Africans, who responded very well when such payments were offered.

All evidence shows that men gave blood overwhelmingly more than women, with little difference between regions and until recently at all times. Younger donors also gave in higher numbers in Africa than in Europe and America. Both these features were partly the result of the early, organized recruitment at military barracks, prisons, and schools by transfusion services that could store blood. As time went by and disease risk rose, especially for HIV/AIDS, donors in the military and in prisons were excluded and the importance of students rose accordingly. Although there were more boys in schools, the girls that were enrolled provided by far the largest source of female donors in Africa.

Among the other consequences of practicing transfusion in poor countries were the frequent shortages of supplies (bottles, bags, and blood-drawing sets), plus the generally more precarious nature of health, as well as the lack of sufficient infrastructure, such as widely available and reliable electricity. Just to take one of the many consequences, these conditions made it difficult to maintain adequate supplies of stored blood outside the big cities in Africa. The widespread response was to recruit donors for immediate transfusion. This was similar to what developed in Europe and America between the world wars, when the local Red Cross branches in Britain and the transfusion services for hospitals on the continent established lists (or panels) of pretested and willing donors who could be called when needed to give blood. In Africa, when the supply of blood was insufficient, hospitals encouraged patients to bring in family and friends who in many cases were paid to give blood by the family of the patient whether known or not by the hospitals. This occurred especially with transfusion for maternity and pediatric cases in Africa, because family members were much more likely to have a strong sense of obligation to provide the blood required. This circumstance also gave rise to the practice of "replacement" donation, whereby families were told to provide blood donations equal to or in excess of that planned for their family members' transfusion.

The best way overall to characterize the historical response to the need for blood donation in Africa was that it was very flexible. Most notably, hospitals were innovative in how they adapted to their circumstances, and the resulting amount of blood secured generally met the ability to give transfusions. This lack of resistance by Africans was not expected and defied predictions of irrational opposition. One way to interpret this response is that blood was one medicine that Africans possessed in the same amount and with the same control as anywhere else in the world. There were no drug companies or expensive chemical manufacturing or rare materials involved. As far as donors were concerned, the history of blood transfusion offers a good example of Africans' ability to organize well and adapt their health care to their needs when the materials were available to them.

Early Sources of Blood Donation

Blood donors were initially recruited by or served the hospitals where transfusions were given, and thus they were part of the original hospital-based transfusion system. This was similar to the way that hospitals found donors in Europe and America in the 1920s and 1930s. Not surprisingly, donors were first sought within the hospital itself, as when Émile Lejeune transfused his military officer in 1918 in East Africa and drew blood from another patient (see chapter 1).[3] An early experiment in the Hôpital de l'étoile in Katanga Province of the Belgian Congo treated 238 African patients suffering from pneumonia with blood drawn from recovering patients.[4] Likewise, the 1940 report of Joseph Lambillon at Katana Hospital in Kivu Province of eastern Belgian Congo also used the blood taken from recovering patients.[5] In fact, a main conclusion of Lambillon was, "in the colonial setting blood transfusion is very easily done, thanks to the large number of chronic patients that are in all the native hospitals who can serve as donors."[6]

That this was not always "voluntary" is indicated by the report of Jean Langeron, a doctor at the Hôpital de l'étoile in Katanga who inspired the study of serotherapy for pneumonia patients. Langeron initially tested the procedure on forty-five patients, mixing the blood of three at a time in order to obtain an "active, therapeutic serum." He reported that the "asceptic drawing of blood is easy to do." A needle was inserted in a vein below the elbow, where "one can without inconvenience draw 500 cc of blood at a time from an adult; we have often repeated this bleeding 5 to 8 days later, without delaying the convalescence of the patient." Alluding to humoral balance bloodletting, the article concluded, "we do not want to propose that it [taking blood] facilitates it [convalescence], even though this is our impression."[7] Another example of early experimentation, albeit with a less aggressive approach to obtaining blood, was reported in Katanga by the obstetrician George Valcke in 1934. He used autotransfusion for an African woman who had hemorrhaged after giving birth by collecting and filtering her blood, then transfusing it back to her.[8]

There are conflicting reports and mixed results of attempts to obtain larger quantities of blood during the Second World War in Africa. For example, a report of the study of African troops in Kenya at the end of the war began, "Considerable difficulty has been experienced in getting voluntary East African blood donors." Evidence of this was offered from the observation that when blood donors were requested from a patient's army unit, "they are invariably supplied by the Company Commander." But the voluntary nature of the donation by those soldiers was questionable. The report stated, "the *askari* [soldier] doesn't seem to know what is happening until he is bled." Whether agreeable or not, "no direct refusals have been encountered." Presumably, the report concluded, this was because of an "*esprit de corps.*" An even more questionable practice was cited of telling noninfective, recovering venereal disease patients that bleeding was "part of the treatment, or as a test of cure."[9]

To explain why there were difficulties in obtaining blood, the report described a study of two thousand troops, selected from three groups ranked by their "intelligence"—taking into account their ability to speak English and whether individuals were tradesmen or came from the lower classes. The selection methods reveal much about British prejudices, but the results also provide some indication about Africans' views of blood transfusion in 1945. For example, initially those in the study were given a simple lecture on the facts of transfusion that stressed safety, the promise of tea, cigarettes, and a donor badge afterward, plus the assurance that the blood would be used for treating Africans only. This was followed by a demonstration of blood taken from a European and a call for volunteers. Only ten of the two thousand soldiers agreed, and according to the report "all belonged to intelligence group A," the upper class, English speakers. When asked who would give blood to their own tribe, only an additional five agreed, all from the two "lower intelligence" groups.

The rest of the study relied on answers to questions, and when initial, poorly phrased ones such as "Are you afraid of venipuncture?" or "Are there any tribal superstitions that prevent you giving blood?" elicited no response, extended interviews eventually revealed what appear to be more informative responses. Thus, 15 percent complained that they would not give blood because of poor army conditions (e.g., bad food). A larger group (31 percent) admitted perceptively that their blood was "bad," due to bilharzia, dysentery, malaria, hookworm, or syphilis. The most common reason for not giving blood (54 percent) according to the report, came "chiefly from the low intelligence groups," who feared their pint of donated blood would be "irrevocably lost." The study offered the following as a typical rationale: "I was born with a certain amount of blood and I don't want to lose any now."

The French in West and North Africa had a different experience with African blood donors during the Second World War. When Gaston Ouary went to Algiers from Senegal in 1944 to study the latest transfusion techniques with Edmond Benhamou, he had already done numerous transfusions in Dakar, which he likened to "a minor surgical intervention." He noted a bigger problem was finding adequate donors,[10] and this was despite the fact that in Dakar there was compensation given to those who donated blood. His explanation, like that of the British in East Africa, was telling of European prejudices as well as of African attitudes about donation. The psychology against giving blood had a number of causes, according to Ouary: "First the habitual indifference of natives to the suffering of others; then the physical fear, the fear of being weakened by the loss of blood; the mystical fear because transfusion is a way of trafficking bloods; and more than a few believe in contracting the disease of someone lying next to them receiving their blood. These, more than religious reasons, are the true causes of the lack of donors."[11]

In contrast, the transfusion center where he trained in Algiers was operated on a completely different and much larger scale. Simplified procedures plus storage of blood through the drying and refrigeration of plasma allowed the collection of blood for later use and the ability to do many more transfusions. The challenge was, therefore, to find sufficient donors, and Ouary was surprised at the success, especially given his views on "native" psychology. He attributed that success to a combination of efforts, including daily propaganda by the press; the offer of supplementary ration coupons for meat, sugar, and cooking oil; and the leadership of the army, businesses, and government ministries, all of whom urged their subordinates to donate and who set examples by donating their own blood.[12]

Ultimately, the large amount of blood required for Allied troops fighting in North Africa and Italy depended on blood being donated by the mass of the population, and here Ouary admitted the most surprising part of the success.

> As for North African natives, they also donate their blood. This was not without difficulty raised at first by religious problems. But they were smoothed out when the Grand Mufti of Algiers approved transfusion and encouraged Moslems to give their blood. The manner in which the peasants of Medjerda and Kabylie [regions in northern Algeria] have responded to the requests of the mobile teams from the Center has been particularly remarkable. It is they who furnish the major part of blood collected by these teams.[13]

Benhamou's plan was to collect and process blood at the Pasteur Institute of Dakar for shipment to Algiers and then use at the front.[14] Because that program began only at the end of February 1945, it never achieved its anticipated scale before the war ended. Still, in 1945 there were 1,236 donors, from whom 142 liters of plasma were shipped to Algiers. More significant in the long run was that 56 liters of plasma and 25 liters of whole blood were kept for use by hospitals in Dakar.[15] As was indicated by both the minister of colonies and Ouary before he left for Algiers in 1944, the colonial administration was eager to expand transfusion in the African colonies. So the Pasteur Institute continued to function as a blood collection center after the war, although on a lesser scale. In 1948 there were 901 donors (of whom, 790 Africans), who gave 71 liters combined of plasma and whole blood.[16]

In 1950 the French colonial administration finally decided to create a modern blood transfusion center to serve not only Dakar and Senegal, but all the French West African colonies and Cameroon. To encourage the great increase in donation that was necessary, especially from Africans, a decree provided for 500 francs for each donation of 300 to 350 cc of blood, plus 100 francs' worth of provisions, which included:

-a half kilo of bread
-a can of sardines [alternatively, 200 grams of cheese or meat]
-1/2 can of condensed milk
-fruit (according to availability at local market)

-a cup of coffee with sugar
-a packet of cigarettes
-3 kola nuts[17]

Whether motivated by these provisions or not, the new Centre fédérale de transfusion sanguine, built on the outskirts of Dakar, registered a dramatic increase in blood donation. The partial year of operation in 1950 saw 268 liters of plasma and whole blood collected, and 896 liters in 1951.[18] In an article published that same year, Jacques Linhard, who became the first director, described the donors as "voluntary," but added, "they are given a compensation [*gratification*] and a snack at each drawing of blood." He acknowledged that the donors included some European soldiers, many of whom had already donated in France or Indochina. There were also a few European civilians but virtually no women donors. Linhard explained that the great majority of donors were Africans: 1,484 out of a total of 1,939 in the first six months.[19]

Thus, this first and largest centralized blood collection service in sub-Saharan Africa used compensation to assure adequate blood donations. The advantages of centralization included the ability to insure quality of screening, selection, testing, and accountability in the process. But the system was doomed by high cost and ultimately the independence of other colonies (see chapter 3). Even before 1960 hospitals in Togo and the Ivory Coast started collecting blood locally for transfusion at much lower cost.[20] The Dakar center continued to serve the newly independent Senegal, thus providing a model for other countries as well as large hospitals in urban areas, using both paid and volunteer donors.

Early Blood Donors: The British Red Cross in Africa

This use of "compensation" followed the custom begun in France during the 1920s, in the era before blood storage was possible, and Parisian hospitals offered a payment for the inconvenience of donors who were on call to give blood when needed.[21] In contrast, at the same time, the British developed the practice of uncompensated donation that became the hallmark of the worldwide Red Cross blood movement, including British Africa during colonial rule. The results were better than had been achieved by the British army in East Africa during the Second World War, but there was great variation depending on locale.

In chapter 1 I described the response to the initial interest in establishing British Red Cross Society (BRCS) branches, especially in the British settler colonies as early as the 1930s. Blood donor panels were also seen as a particularly good way of recruiting Red Cross members. Involvement of branches in blood donations continued to a certain extent after the Second World War. In

Basutoland, for example, a Red Cross branch was formally approved in December 1950 after a visit by a Miss Borley from the London office. Her report stated that Red Cross subscribers had "a strong wish in this district to start a panel of blood donors, and names are already being collected."[22] Similarly, Lady Limerick, head of the BRCS overseas branches, visited Gambia in 1954, a few years after the branch there was formed. She was disappointed at the lack of interest in transfusion by the colony's director of medical services, and in her report back to London suggested what was needed was "help with a Blood Transfusion Service—there isn't such a service in the colony at the moment, but I think they will have to start one soon, and then the Red Cross really could come into its own."[23]

The big change after 1945 in the British African colonies (as elsewhere) was the growth of new hospitals and the staffing of medical personnel who recognized the need for transfusions and blood donors. The increased use of transfusion meant that the old system of donors who were available on demand was increasingly unable to respond to the growing needs. To be sure, on some occasions the old system worked, but on others it did not. A report for May 1954 from the Lagos and Colony Red Cross branch described several emergency calls. One in early May came on such short notice, unfortunately, that by the time volunteers were found the patient had died. On the brighter side, the same report described another call for a rare blood group that found two men at the military hospital who were compatible.[24] Later that year at an Enugu, Nigeria, divisional meeting, a blood transfusion leader reported "a most satisfactory situation, the donors continue to increase in numbers." An example was one "bush mother" who was given three pints of blood at the hospital, and when it was explained to her husband, he asked to be able to thank the hospital. This was arranged, and as a result of his urging further donors were obtained.[25]

One of the innovations adopted to increase blood donations was the appeal to large organizations with many members. These groups included business enterprises, but the most donors were obtained from three other institutions: the military, prisons, and schools. There are many examples of what soon came to be a standard practice, as in Northern Rhodesia, where a 1952 Red Cross report indicated that the African blood bank went through a "rather lean period." In response, efforts were successfully made to recruit "members of the Police Force, the Prisons, students from Munali [secondary school] and volunteers from the compounds."[26] A similar appeal from the general hospital in Lagos was made when a blood bank was organized there in 1953. After the initial campaign for donors, over four hundred men and women were registered from

Ikeja airport, the Nigerian police, workers at the military hospital, the police college, the Nigerian railway, the postal school, and the Kingsway Store.[27]

The Lagos and Ibadan Red Cross divisions in Nigeria made frequent use of new propaganda techniques, such as posters and films, to encourage donation. The LRCS in Geneva supplied the film *Miracle Fluid,* which was shown in 1954 at the police college, to airport staff, and at high schools, the Post Telegraph School and Kingsway Store in Lagos. Audiences were large, and over two hundred volunteers were enrolled. In Ibadan the film was shown at police barracks and the Anglican diocesan training college.[28] *Miracle Fluid* was originally produced by the Canadian Red Cross and required interpretation and translation for most African audiences, who may have been as fascinated by some winter scenes as by the medical technology.[29] The Kenya Red Cross branch reported that a showing of this film up-country, in Nakuru, was "too technical for the African and his ability to understand." Local divisions did the best they could. In 1953 the Ibadan Red Cross reported the practice of showing the film twice, first with sound, then afterward silently with a commentary by a doctor. In March 1954 the Lagos division showed the film five times in ten days and enrolled two hundred volunteers.[30]

Anticipating this problem of cross-cultural translation, the Uganda Red Cross decided in 1951 to produce its own film, the script of which was written by Rev. Taylor of Mukono. The film was also to be the source of filmstrips and still photos. As late as 1956 the Uganda Red Cross branch reported adding footage and commentary to the film. Translations were planned into Luganda and Hindustani. The 1958 Uganda Red Cross report indicated the division in Teso, Uganda, used the showing of the film several times in an effort to prepare the way for a new blood bank the following year. In 1961 the Nyasaland Red Cross branch purchased the Ugandan film with plans to prepare a commentary in the local language. The following year the Zanzibar committee of the Red Cross borrowed the Ugandan film and showed it all around the island.[31]

The success in obtaining blood donated from upper-level school students caused concern in some colonies when the numbers threatened to make the hospitals overly dependent on this one source. A 1953 a BRCS fieldworker, Teresa Spens, reported after a visit to the Uganda Blood Transfusion Service, "I was impressed by the large number of African blood donors." She noted, however, that they were mostly the senior students at schools and colleges and she thought this was not satisfactory, presumably to be so dependent on such a small and young group, although medical authorities did not seem worried. As they explained, "students in residential institutions are often, owing to good food, better able to give blood than other sections of the community." An effort was being made, she concluded, to obtain donors in older age groups.[32]

A different concern, simple lack of donors, prompted the new Lagos Island Maternity Hospital to institute a requirement to insure adequate blood, according to Lady Limerick, who visited there on her tour of West Africa in 1961. "They get their supplies by demanding that each patient should produce three relatives as blood donors." She reported that husbands were lining up to give their pints as expectant mothers were preparing to give birth. This early example of the "replacement" policy became more widespread as hospitals made arrangements for adequate blood supplies, using methods beyond the British Red Cross model of altruistic donors on lists or giving blood anonymously for the good of the whole.[33]

There were limits to innovations, and one place where the British Red Cross clearly drew the line was direct payment for blood donation. A striking example occurred in Nyasaland (present-day Malawi), where the BRCS field officer, Patricia Jephson, signaled a serious problem in her February 1961 report.

> This is very worrying. The African Blood Bank is empty. The surgeon at the hospital has told me he is unable to operate for lack of blood. There is a separate bank for Europeans and European blood is not used for Africans. The Africans are not coming forward at all. There is some political or superstitious reason behind this which I cannot find out. I have appealed for donors in the paper without success. The Hospital still pays five shillings to each donor, but I have asked in this case that Red Cross record cards should not be given.[34]

In June the Nyasaland branch meeting agreed with her recommendation to withhold Red Cross certificates from those who were paid five shillings. When the district medical officer stressed how much he needed Red Cross support, the branch agreed to send additional donors and bring in an experienced blood recruitment organizer. Jephson repeated that the Red Cross could be involved only in a voluntary service, but the district medical service replied that five shillings "was cheap compared to some countries." This was prophetic of what was to come after independence.[35]

One final example of donors in a British colony that portended future developments was the transfusion service established at the new University College Hospital in Ibadan in 1957. The importance of that service in the context of who donated blood was the ability of the hospital to mobilize over five thousand donations annually, whereas in previous years the numbers had been much smaller. The first head of donor recruitment was Una Maclean, who had a medical degree from Edinburgh and accompanied her husband, Peter Cockshott, to Ibadan, where he was put in charge of radiology at the new hospital.[36] Maclean had worked while in medical school at the Scottish National Blood Transfusion Service, so when those in charge of the hospital realized they needed to set up a transfusion unit (a surprise, according to hematologist Alan Fleming),[37] Maclean agreed.

Ibadan was a large city, estimated at the time to have over half a million inhabitants. According to Maclean, before 1957 the small hospital that served the city had no blood bank, so without the ability to store blood, it required patients to secure a donor from family or friends. The new Ibadan hospital followed the Red Cross model of recruiting voluntary donors to supply a blood bank for use as needed. As a result, it employed consultants who produced posters and films, while Maclean organized lectures and mobile blood collections. These collections were done because the new hospital needed as much as one hundred donations of blood per week, as well as the storage facilities to accommodate it. Her description in 1958 of a collection was quite graphic: "A bus loaded with equipment and staff goes every week now to a pre-arranged place, often a school room, and there the sight of other people giving blood does not repel the numerous onlookers as might be feared, but on the contrary, produces so many eager donors that equipment may run out before the supply of volunteers."[38] Despite these efforts, the hospital still relied heavily on patients to provide a significant proportion of donors. As she recalled later in a memoir, there were always two categories of donors: "on the one hand members of captive groups of students, soldiers and senior school pupils, on the other, the relatives of patients who had received a blood transfusion."[39] In 1958, after one year of the new hospital's operation, Mclean reported that the hospital had collected 4,637 pints of blood, but the percentage of "voluntary" donors was only a little over half (52 percent).[40]

As new and modern as the Ibadan UCH was, it followed the hospital-based model of obtaining blood for transfusion. Although the hospital made every effort to follow the voluntary-donation model, it was necessary to make up the difference by other means. After independence, when ties were broken to the British Red Cross, hospitals increasingly relied on family or replacement donation, without much scrutiny of whether payment or other compensation was arranged. Hence, by the end of the 1970s Nigeria had expanded the use of blood transfusions but without a central service to meet demands, and with hospitals using any means possible to obtain blood to meet patient needs. A 2002 study of transfusion at the University of Benin Teaching Hospital reported, "Commercial blood donation accounted for 95.3%, compared to 4.7% from replacement and volunteer donors."[41]

The Postindependence Shift to Hospital-Based and Family or Replacement Donation

After independence the adoption of these innovations increased the ability to obtain blood donors in sub-Saharan Africa. As the use of transfusion grew along with new hospitals, the central coordination, including monitoring and control of blood supply, became more and more difficult. The end of the multi-colony responsibilities of the blood center at Dakar is the most spectacular example of this problem. The colonies and newly independent governments of francophone West Africa wanted the independence and savings of collecting blood locally, and they rarely invested the funds for a new national system.

In anglophone Africa and the Belgian Congo, the colonial Red Cross branches were unable to continue blood collection when they became national Red Cross societies after independence. With only a few exceptions the role of the Red Cross was soon relegated primarily to encouraging blood donation, with hospitals doing the actual drawing of blood. Thus, although the Red Cross donors were unpaid, the hospitals were free to make other arrangements for collecting more blood as needed. In the case of the Kenya Red Cross, already in 1960, four years before independence, the collection of blood in Nairobi was turned over to a Nairobi coordinating committee based at King George VI Hospital (see chapter 2). Even in Uganda, which was probably the most successful Red Cross blood collection program in sub-Saharan Africa, the organization collected blood for only a few years longer than its Kenyan counterpart. In 1962 the government's Uganda Blood Transfusion Service was established when the new Mulago Hospital opened, on the eve of independence. The Uganda Red Cross then shifted its activity toward recruitment and the supply of equipment. Hospitals, especially outside Kampala, had even more leeway in obtaining blood by other means than they had earlier (see chapter 3).

The transfusion centers at big hospitals in large urban settings were able to meet their growing needs (including often the supply of blood to other hospitals in the region) because they had the ability to obtain and store more blood, collected increasingly from soldiers, prisoners, and school children. The King George VI hospital in Nairobi sometimes received surplus blood from outlying district and regional hospitals, as well as replacement donations from families above and beyond what was needed for their relatives. There is not much evidence, at least initially, that the latter was a significant source of blood.

More significant was expansion of the practice of mobile collection, which required a vehicle to transport personnel and blood collected at remote sites such as prisons, barracks, and schools.

From its start, in 1948, the Red Cross Service in Kampala collected blood at remote sites thanks to funds from the London BRCS for a van to transport personnel and blood collected around the city.[42] The importance of mobile collection at Ibadan has also been described, and in Ouagadougou (capital of Burkina Faso) the city's main hospital organized a blood bank that permitted it to expand transfusions from four liters of blood used per week that was flown in from Dakar in 1958, to eighty-six liters per month by 1968. Thanks to the gift in 1971 of a motorcycle from the Fédération internationale des donneurs de sang, the hospital was able to increase the number of donors by providing them transportation.[43] A 1972 newspaper article in Kenya described the collection of four hundred pints of blood at the Armed Forces Training College, near Nakuru in the Rift Valley, with part of the blood being sent to Nairobi and part going to Nakuru Hospital.[44]

An example from the height of success in blood transfusion in Uganda before the collapse in the mid-1970s shows how a decentralized Red Cross worked with local hospitals to recruit donors wherever they could be found. The 1970 annual report of the Uganda Red Cross Society, the Red Cross division at Mubende, in central Uganda, halfway between Kampala and Ft. Portal, provided statistics about donors (see table 5.1).[45] Mubende offers an extreme example of Maclean's observation that donors in Ibadan came from essentially two "captive groups," prisoners and soldiers (almost 90 percent in the case of Mubende in 1970), plus relatives.

TABLE 5.1. **Donations, regional transfusion center, Mubende Division, Uganda, 1970**

Source	Pints
Kaweri prison	138
3rd Bn., Uganda army	106
Muyinayina prison	70
Kiboga prison	54
Magala prison	40
Grail school	15
Technical school	10
Mityana prison	8
Mwera prison	15
Relatives and friends	51
Total	507

Source: Uganda Red Cross Society, annual report, 1970, 31, Periodicals: Uganda, box 16723, IFRC.

Many newly independent countries tried to make blood donation an act of patriotism, although it is difficult to assess their success. One standard publicity event, when blood was in short supply or especially at the opening of new facilities, was the ministers of health or even the presidents being photographed donating blood. At the opening of the new blood transfusion center in Bujumbura, Burundi, in 1972, not only did the head of the Burundi Red Cross donate blood but also diplomats from West Germany and the Soviet Union who attended the building inauguration.[46] The oldest and best-known blood donor event in Africa by far, however, was established in Kenya: an appeal for blood donation the week before Kenyatta Day, each October. Newspapers used the occasion to add human interest stories to the pictures of government officials donating blood. "Girls Lead Blood Drive in Mombasa," was the headline of one story about twelve girls from the government's Coast Secretarial College who donated blood at Coast Province General Hospital, saying, "If the men can do it, so can we." For the 1972 Kenyatta Day the *Daily Nation* aimed more at education with the headline "Blood . . . and Why You Need It."[47]

In the end, and especially in hard times, hospitals were the institutions of last resort that were responsible for collecting blood in most African countries after independence. Given the lifesaving nature of transfusion, they therefore used all means possible to obtain blood, including voluntary, mobile collection and indirect pressure from military commanding officers, prison wardens, or teachers. When necessary (and if blood storage was possible) hospitals required replacement of blood by the patient's family, either from family members themselves or "friends" at double or triple the amount needed by the patient. The extreme case mentioned in an earlier chapter about Mulago Hospital in Kampala, Uganda, during the brutal dictatorship of Idi Amin and the ensuing civil war shows what hospitals did in desperate times. In 1977, Mulago Hospital received 5,849 units of blood from the Uganda Blood Transfusion Service at Nakasero, and it collected 3,493 units for transfusion from family members. In 1984 at the height of civil war, when the new LRCS blood director, Juhani Leikola, visited Kampala, he found that between July and September of that year only 9.5 percent of blood for transfusion at Mulago came from Nakasero.[48]

A more typical example of how blood was obtained for transfusion by a district hospital on a daily basis, month after month, year after year can be gleaned from the following description of transfusions at Katana Hospital in Kivu Province of Congo (formerly Zaire). As was seen in both Kenya and Uganda, after the 1960s there was much faster growth of transfusions in new and expanded hospitals outside the capitals, such that by the 1970s the majority of transfusions were done there. Myriam Malengreau, who worked at Katana Hospital in the 1970s and 1980s, recollected,

> I worked from October 1974 to June 1989, first as a volunteer, then as Belgian technical expert, placed at the disposal of FOMULAC (Fondation médicale de l'Université de Louvain en Afrique centrale), who owned the hospital. . . .
>
> There was no blood bank at Kivu (former province now divided into three: Nord-Kivu, Sud-Kivu and Maniema), and only those hospitals enjoying outside support, such as the one at Katana, which received bilateral Belgian support until 1991, regularly practiced transfusions. Katana had about 10,000 patients annually, 5,000 births, and 600–1,200 major and minor procedures.
>
> In Congo between 1970 and 1990, with the exceptions of Kinshasa and Lubumbashi (to be verified), those hospitals with enough means to obtain equipment for drawing and testing blood organized all transfusions in the same way, on a small scale, by drawing blood from the families of patients, as needed, or with a small stock in reserve, as we had in Katana. No hospital or laboratory supplied blood to patients or to an outside facility.
>
> To insure a permanent emergency supply of blood, when a patient received a transfusion, a family member was supposed to "reimburse" the hospital by giving blood, in addition to paying for supplies (blood bag or bottle and kit with a filter and needle) and tests which could run about 3–4 euros in the 1980s.
>
> The "reimbursement" (in blood) for transfusions always required a lot of perseverance on the part of the staff, because the families were generally reticent, even when it concerned saving the life of one of their own, and even more so when they were already saved . . . or dead.
>
> Sometimes, in emergency cases when a patient lacked a potential donor (often the case in pediatrics because the mothers were pregnant) and when the reserve was exhausted, we donated our own blood and solicited the hospital personnel and nursing students, but they were never very enthusiastic and it required a strong dose of persuasion. We carefully kept in reserve the personnel and students with rare RH negative blood.
>
> Some years, in order to guarantee a supply of blood, we required a donation of blood in advance for every surgical operation, even those which were benign (i.e., independent of the actual individual need). Some returned to claim their blood because the patient did not need it. This system was difficult to maintain and was only partially implemented.

Another problem was to monitor those who came to give blood too often, for financial reasons. Families of patients paid them, we understood, for coming to donate blood in their place.[49]

The 1970s and After: Development of Centralized, "Voluntary" Donation

The mid-1970s saw the leveling off, and in some places decline, in the use of transfusions in sub-Saharan Africa for reasons mentioned earlier. The growth in hospital construction, and accompanying services like transfusions, had continued unabated since the Second World War. We have seen that the rising need for donated blood was met through flexible and innovative responses that were essentially hospital based. Thus, if a hospital were located in or close to an urban area, it could receive blood from a central facility that processed and stored blood collected using mobile units for visits, especially to donors at army, prison, and school sites. If those were inadequate or the hospital was in a location inaccessible to these sources, it collected blood itself, usually by a combination of means, including voluntary and replacement donors arranged by the family of the patient.

This was an evolving, flexible system that changed after the 1970s because of new developments that influenced the amount and proportion of blood obtained from different donors. These were the same influences mentioned earlier that affected all transfusion: economic crisis and instability, the AIDS epidemic, and outside assistance. The first of these to affect blood donation was the economic crisis and growing instability in parts of Africa beginning in the 1970s, which limited or cut support for central collection. The crisis thus forced families to become more and more responsible not only for finding donors, but often also obtaining equipment and supplies. Already at the height of transfusion use in 1970, the Ugandan Red Cross reported, "Each year more blood is required and more demand of refreshments, cigarettes increases. This is a problem everywhere and unless we find means and ways of improving this we are bound to lose our blood donors."[50] This decline in resources for donor recruitment was even more critical because it came in the face of rising demand. By 1975 the Uganda Red Cross complained,

> Blood transfusion service all over the country has experienced two major problems which made it not to collect much blood as it was last year compared to the list given here below showing 1974 collection and 1975 [a decline from 33,855 to 24,227 donations]. There has been transport problem which is a countrywide cry in almost all hospitals where they could not provide vehicles to go and organize blood sessions. Some

hospitals fail to find drinks and cigarettes for the blood donors so they fail to hold any sessions.[51]

Ten years later the AIDS epidemic affected the donation of blood in additional ways. In the short run, it resulted in reducing even more the availability of blood donors, some of whom were fearful of the possibility of contracting the disease by donating blood. It was an unwarranted fear but one that was difficult to dispel. According to medical anthropologist Didier Fassin, this fear produced an "abrupt" drop in blood donation at the Centre hospitalier et universitaire [university hospital] in Brazzaville.[52] To a certain extent this drop was mitigated by doctors recommending fewer transfusions because of the risk of contaminated blood. The Nairobi NBTS reported a dramatic drop in donations—from 14, 216 in 1988 to 8,163 in 1989 (the peak had been over 26,000 in 1981). For the first time, this annual report mentioned fear of being tested for HIV as a reason for the decline, but it insisted that lack of resources was the more important cause of the long-term decline.[53]

A more direct impact of the AIDS epidemic on transfusion was the decision to terminate blood collection from military units and prisons because of high seroprevalence. This reduced available blood even faster than any decline in use. Fear of shortages likely delayed decisions by transfusion services to make these changes in collecting practice until relatively late and in a manner that was not highly publicized. In its annual report for 1990, the Nairobi NBTS stressed the importance of schools for collections and mentioned it had stopped collecting from prisons and the army due to a high percentage of those testing positive for HIV. "This decision was verbally and quietly communicated to all provincial Laboratory Technologists. It was done so in order not to arise [*sic*] the suspicion of the press and politicians."[54] The report also cited lack of available transportation and refreshments as reasons for a general decline in donations.

In Rwanda a change was noted in the 1986 annual report of the Inter-Croix-Rouge blood transfusion project financed with outside support. After it was recognized that the AIDS virus was transmitted by transfusion, "the procedure of transfusion was reduced to a strict minimum in the hospitals, with blood being reserved only for otherwise desperate emergencies." The following year, the report admitted, "a certain decline [in blood donors] began in 1985 following the problem of AIDS. As a result we intervened with the medical personnel so that they use transfusion only in cases where it was truly indispensable, until we had the necessary equipment to detect those seropositive for HIV."[55]

In the long run the most important change in transfusion that resulted from the AIDS epidemic was the attention drawn to the blood supply and the need for outside support to make it safer. Thus, for example, the AIDS crisis was the main reason for rebuilding the Uganda blood transfusion service at the end of the 1980s with outside assistance. The result was a shift away from a hospital-based blood collection system toward a centralized one.[56] Outside assistance to African countries had begun even before the World Health Assembly statement on voluntary blood donation and the LRCS policy of assistance to less-developed countries in the mid-1970s (see chapter 3). The examples of Ethiopia and Burundi showed that recruitment of blood donors could be significantly improved, but the change would be limited if the new institutions and practices did not receive continuing government support. Similarly, in Zaire during the 1970s two American doctors helped establish a blood bank with voluntary donors at Mama Yemo Hospital in Kinshasa, only to have it disappear soon after their departure, in 1978. "The blood bank at Mama Yemo Hospital was reduced to playing a role of simple 'poste de transfusion,'" according to a 1983 report.[57] In this context, the Uganda case was unusual because of the amount and long-term commitment of support from the European Union. EU support lasted from 1987 through 2000, and after a short break support from the United States and PEPFAR began in 2004 (see chapter 3).[58]

Rwanda was also able to institute a national blood transfusion system with strong and continuing support primarily from Belgium. Again, that support began before the discovery of AIDS, and in fact the already-established assistance was crucial in mitigating the impact of the epidemic. The Rwanda project had important implications for international donors that are worth further analysis, because they foreshadowed subsequent large-scale and ongoing programs in the years that followed. The initial expenditures were on modernizing the blood collection center in the capital of Kigali. In the first full year (1977) there were 1,922 blood donations reported and an equal number the following year. The plan was to create blood collection and processing centers outside the capital, beginning with Butare, site of the university and the medical school hospital.[59] In 1980 the Butare center received eight hundred donations and by 1982 over fourteen hundred. Two other centers were added in 1983.

Eventually the transfusion center in Butare had more blood donors than Kigali, but both were overshadowed by the use of off-site blood collection at fixed locations, using mobile vans. This turned out to be the key to the success of the service, which eventually exceeded twenty thousand donations per year by 1991. For example, in 1983 only 7,607 of the 14,596 donations were at the four centers; by 1987 that number had declined to fewer than twenty-five hundred of over fourteen thousand donations; and by the early 1990s fewer than one thousand donors gave blood at the centers.[60]

TABLE 5.2. Donations, Rwanda blood transfusion service, 1976–93

Year	*All donations*	*1st-time donors*		*School donors*	
		number	*%*	*number*	*%*
1976	286	—			
1977	1,922				
1978	2,005				
1979	2,326				
1980	6,277				
1981	8,481				
1982	9,406				
1983	14,596				
1984	8,756	5,744	65.6		
1985	8,128				
1986	9,304				
1987	14,026	10,573	75.4	4,404	31.4
1988	19,072	12,265	64.3	—	—
1989	19,141	11,657	60.9	5,899	30.8
1990	18,189	11,815	65.0	5,518	30.3
1991	20,697	13,810	66.7	5,528	26.7
1992	22,245	13,432	60.4	7,348	33.0
1993	21,607	11,140	51.6	9,963	46.1

Note: The categories of 1st-time and school donors are not mutually exclusive, nor the total subset of all donors.

Source: Projet Inter-Croix-Rouge de transfusion sanguine au Rwanda, "Rapport annuel du travail," 1976–93, Internationale Acties 7/23, 7/24, 7/31, 5/14, 5/22, BRC Mechelen.

The sharp rise and decline in all donations from 1982 through 1984 was independent of the appearance of AIDS, which was only first mentioned in the 1986 annual report (see table 5.2). The decline in 1984 was described as largely the result of a cut in budget, although it also coincided with a new directive to shift away from collecting blood at the urban centers, following a WHO advisory to do more collecting in rural areas.[61] The reasoning was twofold: it would gain access to a larger population base and at the same time be a source of blood from donors less likely to have contagious diseases. The advisory came before the beginning of the AIDS crisis in Rwanda, but the recruitment of blood donors in rural areas probably reduced the spread of the disease by transfusion. As the table shows, AIDS likely also hurt efforts at recruitment of new donors in the country when the disease was recognized, in 1985. Reports indicated that 25 percent of blood collected in 1985 was unusable because of contamination.[62]

For later years, the table indicates another feature of donor recruitment, the significant role of older school children as donors. Over 30 percent of blood donations came from the schools in every year except one for which figures are available, and over 46 percent in 1993. This is part of the reason that the overwhelming majority of donors for these years (72–75 percent) were between the ages of eighteen and twenty-five. Not indicated on the chart, but an even more striking demographic feature of donors, was that over 80 percent were male. This had historically been the case in most African countries, although in Rwanda in the 1990s the percentage of female donors slowly rose from 11 percent in 1985 to 20 percent by 1993.[63]

Finally, these figures reveal another strategy of WHO for safe blood collection: the attempt to secure regular, repeat blood donors. This strategy not only had the convenience of increasing the numbers of those who understood the process of donation, but in addition repeat donors were considered to be a safer source of blood, with a motivation and track record of being freer from blood-borne disease than first-time donors.[64] The figures from the chart show little success, however, in diminishing reliance on first-time donors until the last two years of records, 1992 and 1993.

The statistics available from the Rwanda project offer an indication of one other important feature of these centralized transfusion services: the ability to collect data and study results. This capability was relatively uncommon, at least until after 2000 and a new wave of outside funding, such as from PEPFAR, the U.S. program against AIDS. In addition to Nigeria and Congo, where political and logistical reasons kept a centralized system from being established, even countries such as Senegal, Cameroon, and Kenya, three of the larger, more stable, and wealthier countries of sub-Saharan Africa, had surprisingly decentralized, hospital-based blood collection programs through the end of the 1990s. According to data collected in 2002–3, seven of fifteen African countries reported decentralized blood systems, plus there were doubts about the accuracy of two of the countries reporting centralized systems (Democratic Republic of Congo and Senegal).[65] A more recent assessment (2007) reported only six sub-Saharan African countries with "vertical, centralized systems in which voluntary blood donors are recruited, screened and bled by regional centres, and blood is distributed to peripheral hospitals."[66]

Taking Kenya as one example of a country without a centralized system at the end of the twentieth century, initially the Kenya Red Cross started blood collection for transfusion in the late 1940s using volunteer donors. The Ministry of Health appointed a coordinating committee as early as 1960, thus demonstrating an early government attempt at coordinating blood collection. Yet figures from 1999 in Nairobi show that the overwhelming percentage of donors were family or replacement donors, according to the monthly reports.

TABLE 5.3. **Donations, Blood Donor Services, Nairobi, by month, 1999**

Month	*Total units*	*Voluntary and relatives at center*	*Mobile collection*
January	807	768	39
February	868	776	92
March	1,623	941	682
April	774	753	21
May	1,585	913	672
June	1,279	773	506
July & August	1,744	1,679	65
September	1,407	774	633
October	—	—	—
November	980	885	95
December	660	654	6
Total	11,727	8,916 (76%)	2,811

Source: Blood Donor Services, Nairobi, "Monthly Reports," January–December, 1999, KNBTS.

In almost all cases, the mobile collections were at schools. The division between "voluntary" and relatives' blood collected at the blood center is not given, but evidence strongly suggests that the majority were family members. Not only does that fit the practice elsewhere, but in half the cases the label for the category was simply "relatives at the center." Moreover, the numbers are remarkably steady, suggesting that collection followed the predictable demand of the Kenyatta National Hospital for blood to use in transfusions. The periodic increase in blood collected monthly coincided with times that the service could make mobile collections (only four times reported for 1999). This blood was presumably used by other hospitals in Nairobi. Finally, another study done from April 1999 to April 2000 of over ten thousand donors at the two biggest hospitals in Nairobi (including Kenyatta National Hospital) confirms this source of donors. Over 97 percent were "call responsive," the Kenyan phrase for family or replacement donors.[67] The U.S. embassy bombing in Nairobi brought renewed outside attention to the national blood policy and collections system in Kenya, which was funded beginning in 2004 with PEPFAR support to install a national system.

Historically Africans recruited blood donors in a variety of ways in order to meet the growing demand and ability to give blood transfusions. With some significant exceptions, the most important institutions in determining the process, until recently, have been the hospitals. From the beginning, with few exceptions, they found donors as needed, from family, friends, and those willing to be on call. Most important in this process was not whether donors were voluntary or paid, since in Africa notions of obligation and compensation were more complicated than in Europe or North America. Rather it was the size and facilities of hospitals or collection centers that dictated whether blood was used immediately or stored for later use or distributed to other hospitals.

In the national (or regional) transfusion centers like Nakasero in Uganda, or the Dakar center, and the new, large urban hospitals such as Ibadan, there was electricity and refrigeration for storage. These accommodated the collection of a large amount of blood that required systems of donor recruitment that had a number of features different from hospitals where blood was used immediately. Because blood was stored, the patients were unknown to the donors. Hence, it can be characterized as anonymous, an important distinction in motivation. Thus a "debt" was less likely to be incurred in this situation when the donor and recipient did not know one another, whether compensated with money or not.

More obviously, storage allowed collection of more blood for transfusion, but also created the need to establish a system that used a variety of means to secure more donors. This might involve compensation, as in Dakar until the 1980s, but everywhere eventually mobile units were used to collect blood off-site. These soon became regular sources of "captive" donors, to use Una Maclean's phrase (those at institutions such as army barracks, prisons, and schools). Hospitals, both large and small, still needed to secure immediate donors when conditions or supplies of stored blood were low. The solution was use of the so-called walk-in, family, and replacement practices. This was reported in all countries, as smaller hospitals found their supplies from larger centers disrupted, or the urban hospitals saw declining donations due to budget cuts or civil strife.

The AIDS epidemic is best characterized as another, albeit extreme, reason for decline in donors, especially for stored-blood collection, because two of the largest sources of captive donors with high risk of infection were eliminated: prisoners and soldiers. That same crisis, however, brought closer scrutiny in deciding to use transfusion and renewed outside interest in assisting organizing national systems of stored (and anonymously donated) blood. These systems have the advantages of additional flexibility in assuring supplies and maintaining standards of testing, plus the ability to keep records for assessing practices and planning future needs. Time will tell whether governments will provide ongoing and sustainable support.

6 BLOOD TRANSFUSION AND HEALTH RISK BEFORE AND AFTER THE AIDS EPIDEMIC

In the age of the global AIDS pandemic and the current concerns about blood safety, there is little need to justify a more detailed examination of the history of transfusion risk in Africa. Indeed, a crucial part of the initial research that confirmed the existence of HIV/AIDS in Africa was done in the blood banks of hospitals in Kigali, Rwanda, and Kinshasa.[1] Other evidence of the keen awareness of the link between AIDS and transfusion includes the fact that blood donors were one of the earliest groups widely tested for HIV, the first successful measure to limit the spread of the disease.

The impact of AIDS on the practice of transfusion has been referred to frequently in previous chapters, and of all the features of blood transfusion in Africa, the risk of contamination has been by far the most frequently studied.[2] The main focus of this chapter, however, will be the period before the AIDS epidemic in Africa, not only because it is a neglected subject but also in order to put the matter of AIDS and transfusion into broader historical perspective about risk.

The history of transfusion in Africa shows that from the start those doing the procedure were well aware of risk, and this is not surprising given the disease environment. That does not mean there was a great deal that could be done to prevent some disease transmission through transfusion. For example, to this day blood donors are not screened for malaria in countries where the disease is endemic (much of sub-Saharan Africa) because its wide prevalence would mean drastically reducing the number of transfusions, and those transfused would face exposure to the disease in any case.

This approach to malaria reveals what can best be characterized as a general underlying feature of the history of risk and transfusion, namely that doctors weighed risks and benefits in both deciding whether to transfuse as well as whom to secure as donors. If the risk of death without transfusion was great enough, then almost any means of finding a donor was used. Likewise, once a donor was obtained, some tests might not be done (and the chance of transmitting disease tolerated), if delay or not doing a transfusion risked imminent death.

The main reason for the long-standing practice of weighing risk against the benefit associated with transfusion was that historically the procedure was done in Africa in hospitals under the supervision of doctors. Unlike injections, which were done on the street or in unregulated establishments by untrained personnel, transfusions were not reported, nor have allegations or suspicions been found that they were practiced anywhere except in hospitals under the supervision of doctors. The hospitals and dispensaries varied a great deal in the extent and quality of their facilities and resources, but the fact that transfusion was done by Western-trained doctors means they followed the standard practice in Western medicine of examination of the patient and weighing alternatives for treatment.[3]

The Historical Background of Risk in Blood Transfusion

The history of blood transfusion offers many useful lessons for the study of risk generally in health care. It is a procedure that uses one of the fluids of the body ("ein ganz besonderer Saft"—a very special juice—according to Goethe) universally recognized both for the many cultural as well as medicinal values associated with it. The same reasons, however, that made transfusion so effective as a cure—placing a substance into the circulatory system of the body—also made accidents and errors quickly and dramatically apparent. In fact, this is one reason it took so long to develop effective transfusions; when problems occurred, they were very visible and elicited immediate attention. For example, within forty years of William Harvey's publication of his discovery of the circulation of the blood, in 1628, transfusion experiments began with humans that used a variety of liquids such as animal blood and milk. They were quickly abandoned in large part because their negative or ineffective results were immediately observable and these failures hung like a cloud over subsequent transfusion experiments in the eighteenth and nineteenth centuries.[4]

A successful means of transfusing blood from one human to another was discovered at the beginning of the twentieth century that minimized these risks by moving human blood directly from a donor to a patient. This was first done by literally connecting a donor's artery to a patient's vein in order to reduce the risk of clotting.[5] As luck would have it, Karl Landsteiner discovered a new risk at almost the same time (1901), namely the hemolysis of red cells when blood of different blood types was mixed. (He initially observed three types, but a rarer fourth type was soon discovered.) But that did not slow attempts at transfusions. One reason was that Landsteiner's discovery took time to become widely known. In addition, even those aware of what came to be known as ABO blood types attempted transfusions in cases such as severe blood loss or anemia that ran a far greater of risk of harm or death from delaying transfusion than the risk of a reaction to incompatible blood. In the process, a crude but effective practice arose to test for adverse reactions. In Africa during the First World War, a transfusion patient was first given an initial small amount of the donor's blood, or a "probe," then after a few minutes, if there was no reaction, the rest of the transfusion proceeded.[6] During the First World War the problem of testing compatibility was solved by a more rapid and simple test for blood type, and in the 1920s and 1930s it became a standard part of transfusion practice.

Blood groups were only one of the new discoveries at the beginning of the twentieth century with implications for risk in blood transfusion. For example, August von Wassermann's discovery of a test for syphilis in 1906 was quickly adopted to test donors before transfusion. Another risk was discovered after the First World War in association with the use of sodium citrate, which had solved the problem of preventing coagulation of blood, thus making storage possible. Unfortunately, fever and chills sometimes accompanied transfusion with stored blood and delayed its wide use until the early 1930s, when it was discovered that very strict sterilization was required for the storage containers.[7]

Blood typing and screening for syphilis were the most common testing procedures when blood transfusion expanded greatly after the Second World War. As a result, when new diseases such as hepatitis and HIV/AIDS appeared in the blood supply, they posed a great threat to transfusion procedures that by then had a very low expectation of being contaminated.[8] The solution was exclusion of high-risk groups and development of a test for screening before transfusion. In the United States now there are tests for a dozen infectious agents in blood donated for transfusion, at a very significant cost.[9]

The Risks in Blood Transfusion Specific to Africa

It is difficult to make generalizations about perceptions of and corresponding responses to the risks in practicing blood transfusions in an area as large and varied as sub-Saharan Africa, and including facilities ranging from modern hospitals in capital cities with their own blood banks to small regional or local hospitals, where a doctor might draw blood from a relative or friend to give a transfusion to a patient. Despite these differences, there were some common features in the way transfusion was introduced and practiced in sub-Saharan Africa that affected the recognition of risks and the responses to them.

The first feature stems from the oft-mentioned fact that transfusions were practiced in hospitals either by or under the supervision of doctors. As a result, there was from the start a great awareness of the risks in doing blood transfusions; because they were done by doctors trained in Western scientific medical settings, where there was low tolerance of risk. These doctors were, therefore, very conscious of the increased risk of giving blood transfusion in the African setting, with its much greater prevalence of disease compared to the places where they had been trained.

A second feature of risks in blood transfusions in Africa is that steps to mitigate them varied a great deal, and this reflected the big differences in the kinds of hospitals where they were done. For the most part transfusions began in urban hospital centers, according to data from several colonies and countries in West and Central Africa. Soon, because transfusion procedures had become greatly simplified during the first half of the twentieth century, they could be practiced in more hospitals in Africa. For example, the increased ease of practice developed during the Second World War meant smaller hospitals in remote locations could also do transfusions. The problem is that the ability to do these simpler procedures was not matched by the resources and monitoring for regular, standard testing of donors. Although some colonies and countries planned or attempted centralized transfusion services for collecting, testing, and supplying blood (multiple countries, in the case of French West Africa), these services were abandoned in most places after independence as the demand for transfusion grew and resources became strained. Instead, hospital-based blood collection, testing, and, in some cases, storage was practiced. This was an advantage because it allowed a great deal of flexibility, especially in collection of blood and decisions about transfusion use. The disadvantage, however, was that there were quite varied means and practices for testing and safety, and when the latter was lacking (often the case in district and remote hospitals), the risk was greater.

During the period when blood transfusion became widely established in Africa—in the 1950s and 1960s—procedures called for doctors to assess the main risks, recommend the means to detect them, and take precautions for the safety of the patient. In practice these included, first and foremost, blood typing for compatibility, as well as testing for venereal disease. Screening for other diseases such as malaria or hepatitis was much less frequent and varied greatly. Needless to say, all precautions were not followed before every transfusion was given, but in the period of transition from colonial rule to independence, governments had strong reasons to demonstrate effective health care. So resources were made available to maintain and expand hospitals, including transfusion services with facilities for doing some testing of blood for disease. In the 1970s, however, a combination of diminished resources plus growing risks threatened this status quo.

There were several new developments at this time that increased the risk of blood transfusion. First, there was the discovery in 1965 of a new threat from hepatitis B. It was clear that the disease was transmitted by transfusion, but there was uncertainty about its exact nature and prevalence, as well as the challenge to find more resources for detection of the disease.[10] In addition, the economic decline following the world oil crisis of 1974 and increasing unrest within and between African countries hit most government budgets hard and limited or diminished resources that could be devoted to health in general and blood transfusion services in particular. When added to the increased demand for transfusion, it was very difficult to maintain, let alone increase, measures to provide safe blood for transfusion.

One bright spot beginning in the 1970s was increased assistance to African governments for improving blood transfusion services from international organizations such as WHO and the League of Red Cross and Red Crescent Societies. Their efforts were focused primarily on technical assistance and the setting of standards and only secondarily on providing financial support for the improvement of infrastructure. In any case, up to the discovery of HIV/AIDS in sub-Saharan Africa in the mid-1980s, the most frequently mentioned problem was not potential risk from blood transfusion but the inadequacy of blood supplies. In other words, the greatest risk was thought to be that too few transfusions were given. In a 1988 article Juhani Leikola, head of the LRCS blood program, reported that, based on survey responses, in the early 1980s in Africa there were 440 transfusions per hundred thousand, while in the Americas there were 2,880, and in Europe 4,030.[11]

Early Examples of Transfusion Risk and the Response

The main specific concerns about risk in the history of transfusing blood in Africa were, roughly in chronological order, blood type incompatibility, syphilis, malaria, hepatitis, and AIDS. A concern about risk can be found in the earliest report of transfusion by Émile Lejeune during the First World War in East Africa, where he described giving an "anti-anaphylactic injection" of 5 cc of blood from the donor and waiting five minutes before transfusing the rest of the blood.[12] There is little mention of compatibility testing in subsequent reports and publications, except indirectly in comments about obtaining reagents for blood typing. Laboratory reports between the wars, however, show frequent testing for syphilis (originally Wassermann, then the more rapid Kahn test after 1930), which presumably included some for the transfusions that took place in the hospitals.[13]

After the Second World War, more Western-trained doctors staffed the growing number of hospitals in sub-Saharan Africa, and they were especially struck by the more dangerous disease climate. This was a problem for transfusion, but it was not insoluble. The general attitude of practitioners was perhaps best reflected by G. M. Edington, a pathologist who was first stationed in West Africa during the Second World War and returned to work at the Accra Medical Research Institute in the Gold Coast shortly thereafter. "The high incidence of endemic diseases," he admitted in a 1956 article about the selection of blood donors in Accra, combined with a shortage of staff, made it impossible "to emulate the high standards of selection [of donors] reached in more temperate climates."[14] In the end, he concluded, "A blood transfusion service can be safely established in the presence of hyperendemic malaria and yaws. . . . The incidence of reactions in a tropical climate is higher than that recorded in Europe but the major factor is lack of care in cleaning equipment. The major hazard is most probably over enthusiastic transfusion therapy in the anaemic patient." According to other accounts, the risk of transmitting disease from donors was the most frequently mentioned potential problem. The list of these diseases was much the same throughout Africa: malaria, syphilis, and filariasis, plus, depending on location, trypanosomiasis (sleeping sickness) and the newly discovered hepatitis.

Another category of perceived risk, which was little discussed by doctors but definitely affected practice, at least during the period of colonial rule in some parts of Africa, was race mixing. It was clearly behind the practice of giving transfusions to Africans with blood donated by Europeans but not the reverse. In some colonies with large settler populations (e.g., Kenya, Southern Rhodesia), hospitals established segregated blood services. In a pragmatic sense, the practice was necessary to insure the participation of Europeans. This practice was rarely discussed on the record, in part because it was not medical but also probably because it was such a deep-seated presumption unnecessary to justify. One unguarded comment in a British Red Cross report about the newly independent country of Ghana reflected how this practice was rationalized in medical terms. "Because of the possibility of transmitting a form of anemia very prevalent among Ghanaians," stated a British Red Cross visitor in her 1958 report back to London headquarters, "it is necessary to keep a separate list of European volunteer donors."[15]

Turning things around, one might ask whether Africans perceived risk from transfusions of blood from other peoples. The records are not much help, since little was reported about preferences, or refusals, of Africans to receive or donate blood from different African ethnic groups. In other words little or no tribal influence was noted in reports and records of the collection and use of blood in transfusion. There were no separate blood banks, little or no evidence of refusals or difficulties in finding blood because of refusal to donate or receive blood from a different tribe or ethnic group. One reason may have been that the racism of Europeans made them oblivious to the possibility that Africans might be sensitive on this point, both as inferiors to whites as well as being so "similar" to each other. In addition, the novelty of the procedure worked against the understanding or development of a threat that blood transfusion might pose. In other words, to Africans, blood transfused in hospitals that came from storage in a blood bank may have seemed just like any other kind of Western medicine. Finally, in the more common situation where patients were responsible for finding donors, they could exercise any racial or tribal preferences in that selection, and it would not likely have been something the hospital administration would notice.

Racism aside, what was done about risk varied greatly, given the uneven level of resources available. For example, during colonial rule, centralized transfusion services were established in locations such as Dakar, Senegal, and Luanda, Angola, with facilities for testing blood donors that were similar to those in the metropole. Jacques Linhard, a doctor in the French colonial health service in Dakar, was the first director of the Centre fédéral de transfusion sanguine, which initially sent blood to all of French West Africa. It continued to test and screen blood for transfusion in Senegal after independence and into the 1970s as the national blood transfusion service.[16] Linhard, whose career in Senegal began before the Second World War, also served as a consultant in 1974 for WHO, which provided technical advice for a new transfusion service to be created in Gabon.[17] The screening tests that Linhard described when the Centre Fédéral was established, in 1951, included blood typing, testing for hemoglobin level, a thick smear to check for parasites such as malaria, sleeping sickness, and Guinea worms, plus testing for syphilis.[18] In a 1973 article, Linhard and colleagues described six illnesses for which blood donors were screened at the service in Dakar.[19]

TABLE 6.1. **Results of transfusion screening, Senegal blood transfusion service, 1973**

Illness	*Prevalence*
Hemoglobinosis	9%
G6PD deficiency	10%
Malaria	100% (assumed, hence all patients receiving transfusion treated for malaria)
Treponematosis	14%
Trypanosomiasis	rare, but screened
Hepatitis A	9%

Source: J. Linhard, G. Diebolt, and E. Ayité, "Particularités médicales des transfusions sous les tropiques," *Bulletin de la Société Médicale de l'Afrique Noire de la Langue Française* 18* (1973): 293–97.

In 1955, Almerindo Lessa, director of Portuguese overseas transfusion services, reported similar procedures at Luanda; and large hospitals in other colonies initially also had such capabilities in their blood banks and pathology laboratories: University College Hospital in Ibadan, King George VI Hospital (later Kenyatta National Hospital) in Nairobi, the Central Pathology Laboratory in Dar es Salaam (which did tests for the local blood bank), and the hospital maintained by the University of Louvain in Kisantu, Congo, south of Léopoldville.[20] Most of these services continued after independence, although few were able to do collection or testing nationally, as the number of hospitals and transfusions grew. Thereafter, some new blood transfusion laboratories were created, as in Nigeria, Benin (formerly Dahomey), and the Ivory Coast, where a national transfusion service was created on the French model shortly before independence.[21]

The Assessment of Risk in Reports and Workshops of the 1970s

It is difficult to judge the thoroughness and consistency of this testing, even for the large hospitals with good laboratories, let alone at smaller regional and local hospitals. The latter especially gave numerous transfusions with little or no technical capabilities of testing blood, which was often drawn on the spot from family and friends of patients. Given the resources at these locations, the minimum screening most frequently reported was for blood type, accompanied by an interview with the potential donor. A few reports from the early 1970s by consultants working for WHO illustrate these differences. In October 1972, Kenneth Goldsmith, director of the International Blood Group Reference Laboratory in London, visited Kenya, Tanzania, and Burundi for a report to WHO and the LRCS about the blood transfusion services in these countries. At the Central Pathology Laboratory in Dar es Salaam, he found a staff and equipment that followed "an exceptionally high standard."[22] They did ABO and Rh grouping, as well as testing for syphilis and the Australian surface antigen for hepatitis B. This was not the case, however, at a regional hospital he visited in Tanzania where only blood grouping was done.

In Kenya, Goldsmith found that the National Public Health Laboratory in Nairobi also did blood grouping and tests for syphilis and the Australian antigen, but the staff was overworked and the lab was not up to the standards in Dar es Salaam. Moreover, he found instances of blood being sent to Nairobi from as far away as forty miles for testing and return, sometimes without refrigeration. He cited one example of twenty-five bags of blood, instead of samples, that were sent from Kajiado District Hospital, forty miles away, "with the taking needles still attached," and no ice packing. "Two of the bags had been pierced by the needles during transit and none were considered suitable for transfusion." Goldsmith visited Mombasa, where he found another regional hospital where blood was grouped, but "tests for syphilis were only performed when time permitted." There was no testing for hepatitis.[23]

Another example of the differences in resources for screening can be seen in Linhard's report to WHO on his visit to Libreville, Gabon, in February 1974 to study the feasibility of establishing a blood bank there. At the Libreville Hospital Linhard found significant numbers of transfusions done but no system of stored blood in place. The seventy liters drawn each day came as needed from appeals to soldiers or family members. Soldiers had already been tested for blood group and syphilis. Family members were tested for blood type, but not always for syphilis, due to lack of time. Wassermann tests, Linhard found, were done only once a week. No tests were done for hepatitis.[24]

Judging from the lack of problems noted in reports and the published literature, it appears that blood typing was routinely done before transfusion. How well, is a different question. For example, a review of published literature through the 1980s found only two studies of accidents in blood typing in sub-Saharan Africa. One was a 1961 report of only five accidents in a Dakar hospital over a three-year period when thirty-five hundred units of blood were drawn, a number so low it is obviously suspect. A 1973 study at another Dakar hospital, however, found mistyping to be responsible for accidents that occurred in 3.2 percent of units transfused, and 17 percent of patients receiving transfusions. (The rate was higher per patient because they received multiple units.)[25] This discrepancy suggests that the lack of reports was not the result of lack of actual errors. Rather, it is much more likely that blood-typing mistakes were either ignored or masked by other risks from disease. Additional evidence of concern about possible typing errors can be seen in a recommendation made at the First African Blood Transfusion Congress, sponsored by WHO and the LRCS, held in the Ivory Coast in 1977. Among the first technical recommendations was that "the determination of blood group must be made twice," followed by "a final verification of compatibility at the bedside by the person giving the transfusion, either with another ABO test or by a simple cross-matching test."[26] This was the worldwide blood-banking standard, but the priority of this recommendation suggests that it was likely not widely followed in sub-Saharan Africa.

Other tools for screening in small, remote hospitals were preexamination and the interviewing of potential donors, but their effectiveness in minimizing risk was a long way from the practices in large hospitals and transfusion centers. For example, according to a 1973 Uganda Red Cross newsletter, the following "criteria categorically exclud[ed] a person from giving blood":

- having lived with hepatitis patients for the past 6 months
- having suffered from hepatitis
- having suffered from malaria or lived in a malaria zone in the past 2 years
- having suffered from brucella
- having suffered from convulsions
- vaccination or teeth extraction in the last 15 days[27]

If safeguards against risk depended on obtaining accurate answers from observation or questioning African farmers, there are serious questions about how effectively these measures were able to screen risk in transfusion. Instructions from a Nigerian Red Cross registration form from the late 1960s may reflect a more honest picture of efforts to screen at the local or regional level: "The doctor alone is responsible for deciding whether or not a person may be bled. Record any untoward incidents on the form—i.e., name of any donor or visitor fainting or showing adverse reaction; any injury sustained or clothing damaged, etc. So that the matter may be referred to in case of complaint."[28]

The Risks of Specific Diseases

Most of the reports and literature written about the safety of blood transfusion in sub-Saharan Africa analyzed the risks by disease. Based on prevalence of disease and transmission by transfusion, recommendations were given of how to minimize risk, but they generally provided little justification for the expense of extending screening capabilities. Thus, reports on the prevalence of syphilis indicated that it was low (less than 1 percent in Ibadan and Nairobi in 1980–81).[29] Testing was initially done for syphilis and yaws together, and since the prevalence of yaws was often as high as 50 percent, eliminating these potential donors threatened a dramatic reduction in the blood supply. In addition, Edington warned as early as 1956 that yaws was a disease of childhood, and it was not known if it was transmissible by an adult.[30] As a result, other means were used to minimize the risk of transmitting syphilis and yaws: for example, refrigeration of blood at 4 degrees Celsius for four days.[31] This means of prevention was not entirely effective, because the regional and local hospitals who drew blood from family members for immediate transfusion did not usually have refrigerated storage facilities. Nor did they always have the additional safeguard of giving the recipient a preventive injection of penicillin.[32]

This model of balancing risks and finding alternatives to screening donors with highly prevalent disease was also followed in response to malaria, which was widely recognized as "by far the most important parasitic transfusion infection."[33] Although the thick-smear test was questionable and highly variable, when careful studies were done by the best transfusion centers, rates of 15 to 30 percent were reported.[34] Even if testing could have been done widely for malaria, these rates precluded screening of donors because of the effect it would have on the blood supply. Therefore, where testing was done and found to be positive, precautions were taken to exclude the use of blood from donors with malaria only for vulnerable patients, such as pregnant women, infants, and anemic patients.

In cases where blood was not tested for malaria, which according to reports was the majority of cases, it was widely advised to give preventive malaria treatment either in advance to potential donors, or to patients before or immediately after receiving a transfusion.[35] Follow-up studies were rare and masking was common, so there were few studies of malaria from transfusion in Africa.[36] Whether this situation reflects the success of the preventive treatment or tolerance of a risk, it did not slow down the use of transfusion.

Contrary to common perceptions, HIV was not the first new disease that threatened the blood supply. The discovery of the hepatitis B antigen (HB_2AG) by Baruch S. Blumberg in an Australian aborigine in 1965 set off a renewed interest in the risk of disease from transfusion, primarily because of better understanding of hepatitis and the availability of a test.[37] Of note, the first studies in Africa showed a higher prevalence in African countries (around 6 percent) than almost anywhere else in the world.[38]

TABLE 6.2. **Incidence of HBsAg in blood donors, 1978 review**

Place	*No. tested*	*No. positive*	*% positive*
U.S.			
Rochester	—	—	0.06
Chicago	—	—	1.47
W. Europe			
Denmark	10,000	18	0.18
Belgium	6,656		0.15
Paris	18,046	75	0.42
Japan	—	—	1.05
Kenya	1,167	77	6.60
Tanzania	3,236	134	4.10
Senegal	959	122	12.72
Pakistan	1,918	52	2.70
Australia (Sydney)	56,140	63	0.11
Hawaii	—	—	0.12
Pacific Islands	—	—	10.00

Source: Zuzarte and Kasili, "Hepatitis B Antigen."

The more important question was what the blood collection and transfusion system would do about this newly discovered risk. The answer was, very little—largely because of lack of resources, either to test or to compensate for loss of potential donors.[39] In addition, authorities themselves were not exactly sure about the risk of carrying the antigen. For example, many studies described carriers as otherwise healthy.[40] Thus, a familiar pattern followed of the transfusion centers and large hospital blood banks testing and screening for the HB antigen, with the regional and local hospitals left to rely on a physical examination of patients to see if they showed jaundice.[41] As with malaria, there were few follow-up studies to see if transfusion had induced hepatitis, but there were no disasters reported either.

Other potential risks from transfusion were mentioned in reports and published literature, but either their dangers proved minimal or the cost of screening and cure prohibited wide adoption. Thus, filariasis (worms) proved not to be transmissible by blood transfusion. Relapsing fever and sleeping sickness were relatively rare and were detectable, to a certain extent, by interview and observation. The genetic deficiency G6PD found in 10 percent of donors screened in Senegal in 1973 (see table 6.1) was expensive to detect, and its effect was only on certain patients on other medication.[42]

Despite the flurry of studies and concern in the 1970s about hepatitis and other new diseases, a problem more frequently mentioned was the lack of sufficient donors. For example, those attending the First African Blood Transfusion Congress held in 1977 in the Ivory Coast emphasized the need to increase funding for recruiting donors and establish the legal and technical infrastructure for transfusion services. With many countries represented, most but not all either francophone or West African, these common problems were shared by all, especially since the congress was held in the immediate aftermath of the 1970s economic crisis.[43] There was, however, significant discussion of screening for disease, which offers one benchmark for concern about risk on the eve of the AIDS epidemic. Hepatitis B was the subject of much discussion, with the congress attendees admitting that the HB antigen was found with greater frequency in sub-Saharan Africa. Yet it was also noted that "post-transfusional hepatitis appears to be rare," although that did not mean infection was avoided in those thus exposed, because the disease could be a slow-developing infection.

Not surprisingly, there was disagreement about what course of action should be taken: some wanted to exclude all carriers and others did not believe it was worthwhile testing. In the end a compromise was worked out for a short-term recommendation that all could support. Transfusion organizations should be able to detect antigens, and carriers should be excluded except in case of blood shortage. In such instances blood that tested positive from a healthy carrier could be given to a patient who was also a carrier. In the long-term, the congress recommended, a more sensitive test needed to be developed.[44]

As for malaria, the congress recognized that countries where the disease was not endemic were sensitive to the problem it posed for them. On the other hand, the report noted, "the risks are less for most African countries where malaria is endemic and the immune state of adults is generally good." The congress therefore concluded that systematic testing of all blood donors for antibodies from parasite infection should be part of a "future antimalaria campaign in the third world in general and Africa in particular."[45]

Given the disparity of resources both between and within countries, it was unlikely that the recommendations of the 1977 Congress could uniformly be put into practice. At a 1979 workshop in Nairobi on blood transfusion services in developing countries sponsored by WHO and the LRCS, there was even less concern expressed about risk. Rather, there were other, higher priorities that reflected the problems that had arisen since the mid-1970s, mostly concerning resources. The report by WHO/LRCS from that workshop listed the following "main problems":

- inadequacy of blood supply
- lack of an adequate donor recruitment programme with undue reliance on donation by relatives of patients
- lack of motivation of donors and resistance due to taboos, superstition, fear, religious views
- lack of transport facilities
- lack of equipment e.g. blood bank refrigerators
- lack of legislation on donation of blood
- lack of organization of national blood transfusion services
- lack of standardization of procedures and equipment
- lack of trained manpower
- lack of finance

In the entire sixteen-page report, the only reference to risk was one page devoted to reagent preparation for compatibility testing, but that was mostly part of a discussion about the possibility of cutting the costs of buying reagents through the use of local production. Indirect evidence that there were problems in compatibility testing came from another section of the report, on "laboratory procedures in investigation of transfusion reactions." Six of the seven "final recommendations" of the report called for measures to increase the blood supply and improve technical efficiency. Only one concerned blood safety, a rather general admonition that "the technical procedures undertaken by a blood transfusion service should be designed to insure optimal safety of both donors and recipients of blood."[46]

A good measure of the state of testing in the field, at the height of transfusion use and on the eve of the AIDS crisis, comes from the 1980 annual report of Kenya's National Blood Transfusion Service, in Nairobi, the most important blood collection center in the country. This center screened all donors for hepatitis and syphilis (with corresponding infection rates of 1.5 and 4.8 percent).[47]

Table 6.3. **Blood Bank and Kahn test report, selected tests, National Blood Transfusion Service, Nairobi, 1980**

Number of donors grouped	21,878
Blood issued to Kenyatta National Hospital	10,268
Blood issued to district and provincial hospitals	1,194
Blood issued to private hospitals	5,237
Hepatitis-associated antigen tests done	21,878
Positive hepatitis tests	327
VDRL tests done	21,878
Positive VDRL tests	1,044
Expired blood for freeze drying	1,145

Source: "Blood Bank and Kahn Test Report—1980," KNBTS.

Another indication of how screening was done at all levels can be seen from the WHO/LRCS workshop held in Zimbabwe in December 1985, just as the AIDS epidemic appeared on the world scene. The participants at the workshop recommended the following "Donor Selection Criteria" that would bar a blood donation:

- under 17 or over 65 years of age
- under 50 kg
- over 4 (male) or 3 (female) donations annually; and less than 2 months since prior donation
- hospitalization with jaundice
- previous major surgery or transfusion (within 6 months)
- tooth extraction (within 2 weeks)

- immunization with live vaccines (polio and measles within 2 weeks; yellow fever within 6 weeks)
- any serious chronic illness
- pregnancy
- lactation up to one year
- certain transmissible diseases
- blood pressure (diastolic) above 100
- hemoglobin level below 12.5 (females) or 13 (males), although subject to reconsideration[48]

This screening could not be done, especially at smaller hospitals, without the resources of the blood center in Nairobi.

Blood Transfusion after the Discovery of AIDS

The discovery of HIV/AIDS had dramatic, although not unprecedented, repercussions on the practice of blood transfusion. The response to the HIV contamination of the African blood supply was relatively quick and successful, at least compared to efforts at responding to other crises in Africa. One important reason was that the transfusion infrastructure was relatively well developed, especially given the history of awareness of risk and testing of blood. Since screening was primarily determined by available funding, and that was soon found outside Africa, once the tests and resources became available, the hospital transfusion systems were able to use them quickly and effectively.

In fact, the widespread practice of blood transfusion played an important role in the discovery of AIDS in Africa. The first Africans found to have AIDS were in Belgium and France, as reported in several articles published in 1983. The following year the first articles appeared about Africans with AIDS in Kigali, Rwanda, and Kinshasa, Zaire.[49] Once discovered, researchers quickly went to blood banks in Africa to confirm the existence of AIDS. Among other reasons, this was significant because many still suspected the disease to be linked to homosexuality. In February 1985 the first article was published by Philippe van de Perre on blood donors in Rwanda, and the following year Jonathan Mann began publishing his findings on blood donors in Kinshasa, where he had worked since 1984.[50]

The extent of blood transfusion and the numbers of donors provided overwhelming and breathtaking evidence of an 8 to 18 percent infection rate among blood donors in Rwanda and Zaire. The figures from Uganda are even more chilling (see chapter 3). The special envoy from the European Union investigating the blood supply in Kampala estimated a seropositivity rate between

15 and 20 percent in 1987. Monthly reports of HIV testing of donors at Mulago Hospital in 1988 and 1989 rose and fell between 15 and 33 percent. From July through December 1988, records show that of 1,559 potential donors tested 372 (24 percent) were seropositive.[51]

In the broader context, chapter 3 has shown that the number of blood transfusions peaked in sub-Saharan Africa by the end of the 1970s and early 1980s, hence they had already leveled off or were in decline before the impact of AIDS, due to lack of resources and budget cuts from the economic crisis. Nonetheless, when the AIDS epidemic was recognized in Africa, it decreased transfusion use even more, in part from the erroneous fear of contracting the disease by giving blood, but also because of increased caution by doctors about doing unnecessary transfusions. A more dramatic impact that further reduced the supply for those blood services that collected blood for blood banks was the exclusion of two categories of high-risk donors: prisoners and soldiers. In 1987 the WHO global program on AIDS was established, with Jonathan Mann as its director, and he was quick to create the Global Blood Safety Initiative, which recommended "stopping the practice of using soldiers and prisoners as group blood donors and substituting secondary school students."[52]

The elimination of these high-risk donors is one indication of African blood collection officials' awareness of risk when AIDS was discovered. They were also quick to recognize the need for testing. In Kenya, the Ministry of Health issued a memorandum requiring HIV screening for all blood donors in Nairobi beginning in June 1985.[53] Zimbabwe, with the assistance of funding from Finland and medical direction from Jean C. Emmanuel, claimed to have begun routine screening of donors for HIV in August 1985.[54] The authors of *AIDS in the World* (1992) did a survey in 1991 of fifteen countries, nine of which were in Africa. All fifteen reported beginning screening for HIV by 1986, and all but one reported that 100 percent of blood units collected in 1991 had been screened for HIV.[55]

Based on other studies with more reliable data, it appears these reports likely reflect aspirations and awareness of need rather than widespread actual practice. In Nigeria, for example, only one government hospital required screening of donors in 1988. In fact, some hospitals reported that many people volunteered blood who thought they might be seropositive, just to find out. By 1992 the government had mandated testing, but there was no legal enforcement.[56] In Kinshasa in 1988 two of the three blood banks in the city, including Mama Yemo Hospital, were renovated to test for HIV. Yet a spot check of transfusions done in all Kinshasa hospitals in February 1990 showed that although all transfusions were matched for blood type, of the 3,741 units of blood transfused, 1,045 (28 percent) had not been screened for HIV.[57]

Rwanda was one of those countries reporting 100 percent HIV testing, normally a dubious claim, but there is evidence to support that claim, thanks to the program of outside support that had been established well before the AIDS crisis. The Belgian Red Cross led an effort to build a national blood transfusion service in Rwanda following the riots and disruption in the early 1970s. Therefore, by the time AIDS was first diagnosed in Rwanda, at the end of 1983, there were not only multiple collection centers but the means for testing and outside links to the best medical laboratories in the world. For example, the first twenty-six suspected cases of AIDS were confirmed by tests on blood flown to the CDC in Atlanta. The numbers detected grew slowly (forty-seven in 1984, eighty-nine in 1985), but Rwanda was at the center of the focus of attention in Africa and hosted a congress on diseases transmitted by AIDS in January 1986.[58]

Through its strategy of recruitment in rural areas, restrictions on transfusions, and with close technical support from Belgian and WHO advisers, Rwanda was able to establish an extensive screening program. Van de Perre, of Project SIDA, led the latter effort, which began in December 1985. By September 1986 widespread testing of blood, plus more rigorous prescreening of donors, began to have an impact. The seropositive figure for potential donors in 1986 was 9.5 percent, and slowly declined in subsequent years: 1987 (6.8 percent), 1988 (5.4), 1989 (2.8).[59] In the early 1990s the rate was up slightly, to an average of 3.4 percent. The confidence inspired by these results explains, at least in part, why blood donations rebounded so quickly. In 1987 the number of donors was back to the pre-AIDS level, and in 1988 donations and transfusions went well beyond those figures.[60]

An example of ongoing advice and technical support available to all transfusion services in Africa, thanks to networks established in the years before the AIDS crisis, can be seen in a statement issued by WHO and the LRCS in 1990. Among the key guiding principles were that epidemiological studies "necessary for establishing the prevalence of infectious agents in the population . . . should include prospective and retrospective documentation to assess the risk of transmission of infectious agents by transfusion." Realistically, however, the statement recognized that cost effectiveness "takes into account the risk of transmission based on the prevalence of the infectious agent in the population and the likelihood of immunity; the consequences of infection; and the costs of performing the tests."[61] It recommended screening policies that varied greatly for eight specific diseases.

TABLE 6.4. **WHO/LRCS, "Consensus Statement on Screening of Blood Donations for Infectious Agents Transmissible through Blood Transfusion," February 1, 1990**

Disease agent	*Recommendation*
Retroviruses	"Policies for screening blood for HIV before transfusion should be determined at the national level, taking into consideration the availability of resources, [and] the local prevalence. . . . The appropriate technology must be appropriate. . . . Rapid screening tests may be used when blood is needed urgently."
Hepatits B virus	"Each unit of blood or plasma collected should be tested for HBsAg... and only those giving a negative result should be used."
Blood-transmitted non-A non-B hepatitis (NANBH)	"tests are available but expensive. There is an urgent need for development of less expensive screening tests."
Cytomegalovirus	"blood to be used in high-risk recipients should be tested."
Syphilis	"screening for syphilis is less cost-effective than storing blood at 2–8 degrees C for 72 hours before use."
Trypanosomiasis	"few data are as yet available on the specific problem of blood transmission."
Malaria	"at present no adequate test method"
Toxoplasmosis	"universal screening of blood donors is not recommended."

Source: World Health Organization/League of Red Cross Societies, "Consensus Statement on Screening of Blood Donations for Infectious Agents Transmissible through Blood Transfusion, 30 January to 1 February 1990," WHO/LBS 91.1, WHO archives, Geneva.

Forty years of experience with balancing benefits and risks in transfusion provided not only a framework for understanding the risk but also a system to build on when resources became available from international organizations such as the World Bank, WHO, and the Red Cross, as well as bilateral programs such as the U.S. PEPFAR program to reduce risk of HIV/AIDS from blood transfusion in sub-Saharan Africa. Compared to the public health infrastructure to prevent sexually transmitted diseases, African transfusion services were much better able to limit the transmission of AIDS through contaminated blood.

The obvious benefits of transfusion plus the relative ease of use explain its rapid and wide spread in sub-Saharan Africa after the Second World War. There were limits, however, to its growth, the two main ones being numbers of trained personnel and availability of blood. The former was a much more important limit, since only doctors in hospitals did transfusions. As for the question of risk, the participation of doctors in transfusion was key to the heightened perception that helped prevent unnecessary transfusions and encouraged awareness of the need for safe blood.

Lack of blood was the second important limit on the use of transfusion. In fact, the cost of establishing centralized blood-banking systems, plus the lack of sufficient volunteer donors proved too restrictive, and doctors soon found more immediate sources of blood from patients' family and friends. This was a reversion to the earliest days of transfusion, but it permitted far more transfusions than would have been possible using blood from a centralized system. There was an increased risk, however, from the inability to screen blood at a laboratory adequately equipped for testing infectious parasites. This was especially the case when blood was drawn for immediate use from family members or friends. Doctors at small regional or local hospitals simply did not have the resources or time to test blood for such diseases as hepatitis and sometimes even syphilis, as their counterparts did routinely at centralized blood collection and banking centers.[62]

From the start, doctors in Africa were aware of the trade-off of risk versus benefit of transfusion, and they made accommodations accordingly. Blood typing was presumably the most widespread test performed, judging by its universal mention in all accounts of transfusion services, plus a lack of reports about failure to do so. Testing for disease depended on weighing the risk and cost versus benefits.

By far the greatest disease risk in transfusion in Africa was malaria. Because of the high prevalence plus the difficulty of testing, however, it was quickly recognized, even in the large transfusion centers, that the risk was better mitigated by prophylactic treatment rather than screening. This approach was mainly justified by the great benefits of transfusion, which made it worth the risk of malaria transmission. The availability of a treatment, of course, also weighed in the decision, but it must be assumed that not all transfusions, especially at small regional and local hospitals, had resources for the treatment.

The response to the emergence of hepatitis B followed from the experience with other risks from blood transfusion. The high cost of testing meant screening could be done only at the large hospitals and centralized blood transfusion centers. On the other hand, the relatively low prevalence of hepatitis B meant that those testing positive could be screened without jeopardizing the blood supply. Together, the high cost plus relatively low prevalence made this a moot issue, because testing was not widespread outside the capital cities and largest hospitals.

It is not surprising that the blood transfusion practices in sub-Saharan Africa were incapable of detecting the emergence of HIV. Even the most sophisticated screening in advanced health care systems could not do that. Nor did Africans have the resources to immediately improve their ability to detect and screen for the virus once it was recognized. Given the wide use of transfusion, it therefore was an important reason for the spread of AIDS.[63] Yet African health officials were not without means to understand and respond to the new danger, thanks to forty years of experience and a framework of appreciating long-standing health risks. By taking advantage of outside support for testing HIV, as well as the framework of technical advice and training, the risk of disease transmission through the blood supply was quickly controlled thanks to developments going back to the 1960s.

7 AFRICAN BLOOD TRANSFUSION IN THE CONTEXT OF GLOBAL HEALTH

John Keegan has observed that only anesthesia and antibiotics have been the equals of transfusion in saving lives in wartime.[1] The history of transfusion in Africa is therefore best understood by keeping in mind several features of the procedure that have made it one of a handful of the most important therapies discovered in modern times. First among these features is its clearly demonstrable and comprehensible effectiveness. Dramatic resuscitation is almost immediately produced, and the logic of replacing a lost substance of such universally recognizable importance is understandable to patient and donor alike.

In Africa, transfusion was introduced as part of the medicine of the colonial rulers. So it is difficult to say that Africans independently welcomed the procedure. There was resistance or opposition to some medical procedures that were painful or did not show immediate benefit, such as vaccinations or lumbar puncture for sleeping sickness diagnosis. Although blood donation has some features similar to this, the results of transfusion were among the most immediate and dramatic. Hence, it was a main reason for the adoption of the procedure.

A second important feature of blood transfusion was that although the concept of replacing vital body fluids was ancient, actually doing it required the understanding of a number of medical concepts before it was successfully adopted on a wide scale. Despite many attempts from Hippocrates to Harvey, the effective practice of transfusion came almost 250 years after the discovery of the circulation of the blood. Doctors needed, for example, to appreciate antisepsis in order to remove blood from a donor and reintroduce it to a patient safely. It also took a long time and much trial and error before finding that blood itself was the most effective substance for transfusion. Almost simultaneous with that demonstration was the independent discovery of differences between the immunological properties of peoples' blood. Shortly thereafter, the compatibility of blood groups was recognized as linked to safe transfusion, although it took another decade before tests were rapid and accurate enough to become a standard part of transfusion.

This meant that the history of transfusion in Africa was not a simple process of technology transfer. It could not be done without the necessary accompanying infrastructure, essentially found only in hospitals. This was needed both for diagnosis and treatment of conditions requiring transfusion, as well as the technical procedures associated with it, such as blood testing and asepsis, in addition to the actual collection of blood for donation and monitoring patients afterward. Moreover, the transfusion procedures that were first introduced to Africa did not remain stagnant after their initial discovery and successful application, during the First World War. New methods were developed and implemented throughout the twentieth century for such things as storage of blood, equipment for drawing and giving it, and the testing and organization of donors. In addition to new transfusion techniques arising in medical science, conditions within Africa changed dramatically, which had important consequences for the practice of transfusion. Beyond the obvious political and economic changes, there appeared new educational institutions for training doctors, as well as local, national, and worldwide health organizations that had a direct impact on medical practice.

These complexities of transfusion made it likely to be performed by or under the supervision of doctors and in a hospital setting, and there is no evidence to challenge that assumption. The record shows that the introduction of transfusion to Africa, in the 1920s, occurred after several decades of establishing the first hospitals where the new scientific medicine was practiced. The subsequent history of transfusion was greatly influenced by the continuing growth and change in hospital construction and location, as well as by doctors who worked there and guided the modern health system. As a result this history also reflected changes in the production and employment of doctors, initially foreigners, who served in Africa during and after colonial rule. This increase in the number of doctors eventually included the growing number of those educated in the new African medical schools with supplemental foreign training and assistance. African doctors practiced in hospitals whose numbers continued to grow rapidly until the 1970s. The medical schools and hospitals, as well as the doctors, were influenced by the global system of assistance to developing countries that arose after independence. Transfusion services proved to be institutions frequently supported because of their relatively low cost, desirability, and immediate benefit in the eyes of politicians, medical personnel, and patients.

A final essential feature of transfusion that was unusual in the health field was its reliance on a universally available medicine, human blood. Despite numerous attempts, no feasible substitute on a wide scale has been found for whole human blood. This meant, among other things, that the practice of blood transfusion required the cooperation of many other people besides patients and health providers for it to work. Blood donors were essential.

The recruitment of donors faced some unusual difficulties in Africa because of cross-cultural differences in persuading people to give up something so widely recognized as integral to their body. Nor were hospitals or transfusion services able until recently to draw on many historical traditions of self-sacrifice for national or other identity (an exception being Kenyatta Day in Kenya) to inspire blood donation. Unlike most European and other developed countries, Africans had no long-standing Red Cross or wartime traditions that were used to secure blood donation. But through a combination of compensation that proved especially effective in poorer economies, plus appeal to family responsibility, African hospitals showed remarkable flexibility and resourcefulness in obtaining blood for transfusion. Although not fully compliant with WHO guidelines for voluntary donation, or recommended transfusion needs per hospital bed, African countries have continued to use transfusion as an unparalleled lifesaving procedure, even in the face of dramatic economic, political, and epidemic upheavals.

Because transfusion in Africa was integrated into the hospitals and health practices, changes have followed the general course of improvements (or decline) in health care. Unfortunately, one consequence is that the practice of transfusion undoubtedly played a significant role both in the spread of human immunodeficiency viruses, especially before they were discovered, and possibly in the adaptation from their simian predecessors.[2] Multiple instances have been found of adaptation of the millennia-old simian viruses to human immunodeficiency viruses, all after the beginning of the twentieth century (see chapter 1). It is possible that earlier adaptations occurred, given the long interaction between humans and simians as sources of meat from hunting or as pets; but there are no records of these earlier human viruses from retrospective diagnosis of symptoms or antibodies in human remains.[3]

The lack of earlier evidence points to new developments in the twentieth century that permitted multiple adaptations. To the extent that this occurred because of the increased likelihood of transmitting viruses, no new development in the twentieth century permitted the transmission of disease as efficiently as blood transfusion. Mirko Grmek, in his early and insightful *History of AIDS* (1990), called transfusion the "royal road" for microorganisms that had previously had only narrow paths for transmitting infections.[4] Given the wide spread and increasing use of blood transfusion in sub-Saharan Africa, as shown in this study, this otherwise beneficent medical procedure must be considered a central explanation for the adaptation and emergence of the new human immune viruses, especially in epidemic forms. Unlike transmission by sexual activity or unsterile needle use, with rates of infection between 0.04 and 2.4 percent (depending on viral load of infected person), transfusion from infected blood is almost 100 percent.[5] For epidemiologists who want to assess the role of transfusion in the emergence of HIV/AIDS, the findings of this book can add to our knowledge of where and when transfusions were given in Africa, who donated blood, and who was likely to have received transfusions based on use.

One finding is that transfusions were so ubiquitous by the time of the earliest-found HIV-infected sample (1959), let alone when the disease was first recognized (1981), that there is little doubt transfusions helped spread this unknown disease. The confirming studies done in African blood banks found HIV rates from 10 to 30 percent of donors, shortly after the first HIV/AIDS cases were discovered in Africans living in Europe.

As for the more specific question of whether those who donated and received blood were likely to have transmitted HIV infection, this study has shown that until recently men in a fairly narrow age range (eighteen to forty), the most sexually active years, were the overwhelming majority of donors. Men with serious injuries incurred as accident victims (in factories, traveling on roads, and in hunting or farming) also received many of the transfusions.

There are few records indicating the likelihood of a transfusion recipient being a subsequent donor, but one can reason from historical practice. Men with injuries or undergoing surgery, women in maternity wards, and children with anemia were the principal categories of recipients of blood transfusions, and circumstances encouraged adult males receiving blood transfusions to become donors. As a general phenomenon—known as the grateful patient—someone whose life was saved by a transfusion would be more likely to return the favor. That likelihood was increased by the practice of collecting blood from groups such as prisoners, soldiers, and older male school children. Patients in those groups who had received transfusions would be inspirations for others to donate, including themselves once they had recovered. Moreover, there is little evidence of restrictions on multiple donations or screening for illness, other than those obviously manifesting symptoms or underweight.

So far these observations have assumed a fully adapted, epidemically contagious virus. But studies have shown that simian viruses are not zoonotic,[6] hence they had to adapt to become the human viruses that produced the AIDS epidemic. When and how this occurred is by no means agreed.[7] For example, it is not likely that a random variation produced the adaptation because it has been found to have occurred a dozen times. The serial passage theory is one explanation of how adaptation might have taken place with the limited frequency of crossover adaptations. The idea is based on the finding that a person initially infected with a simian virus has only minor symptoms, which are suppressed in a few months. If that person infects someone else, however, the chances of further variation are greatly increased, and with as little as a third passage, a much more highly lethally adapted "human" virus is possible.[8]

Whether transfusion facilitated serial passage depends in part on the likelihood of transfusion recipients becoming donors, as analyzed above. It also depends on whether transfusions were available when and where the human viruses emerged. The earliest and most studied of the human viruses was the HIV-1M virus, which has so far produced about 95 percent of the cases of the pandemic. The simian source of HIV-1M has been located in southeastern Cameroon, and the earliest tissue sample infected with the virus was found in Congo (Kinshasa), dating from 1959. Based on measurements of the rate of subsequent changes in the form of the virus, molecular biologists have been able to project backward to estimate the likely date of their most recent common ancestor to be the 1920s.[9]

This date and the location in the Congo basin are strikingly close to the earliest dates and locations of blood transfusion found in this study. Moreover, the place and estimated times of the earliest forms of the HIV-2 viruses in West Africa (Guinea-Bissau, Ivory Coast, Sierra Leone) are later (the 1940s) and well within the time that blood transfusion was being practiced all over sub-Saharan Africa.[10] Transfusion alone, however, does not provide a simple explanation of adaptation. For one thing, documentation so far does not indicate that transfusions were very frequent in the 1920s and 1930s in Central Africa. In addition, other means of viral transmission were also increasing at this time, such as extramarital sexual relations encouraged by migrant laborers moving to urban and other work centers. Likewise, the unprecedented injection campaigns against disease were another source of the spread of disease between humans to the extent needles were not adequately sterilized, and these programs began in the 1920s and accelerated after 1945.[11]

There are many reasons to think that adaptation to the epidemic form, just as the epidemic itself, was the result of a combination of circumstances that took place over years and maybe decades. This is in contrast to the smoking-gun scenario that is often assumed about epidemics once a new disease appears in a previously untouched population and rapidly spreads, such as smallpox in the sixteenth century among the Aztecs or cholera in nineteenth-century

Europe. If so, then HIV adaptation was likely the result of all these means of transmission acting together, not just to facilitate adaptation but also to spread the HIVs. The relative importance of each mode, including transfusion, requires much more focused historical research on the time and circumstances of where and when the viruses emerged. The records of hospitals in Cameroon, and the former French Congo (present-day Republic of the Congo) and Belgian Congo (present-day Democratic Republic of the Congo), especially outside the capitals, might be invaluable. Moreover, any work on the West African setting would add to the very scant historical knowledge of population movements and disease campaigns, let alone transfusions in Guinea-Bissau, the Ivory Coast, and Sierra Leone.

Transfusion is not a cutting-edge therapy in modern medicine, so barring unforeseen circumstances there is not likely to be additional outside assistance for improving transfusion in Africa, as is the case currently with the campaigns against infectious diseases. The AIDS epidemic was an exception that brought some additional resources to most transfusion services that would otherwise not have been the case. Most of the twelve countries in sub-Saharan Africa receiving PEPFAR assistance have significantly improved transfusion services and in the process have raised the standards of transfusion for other countries to emulate. For them changes will most likely come from general improvements in hospital care in a given country, which in turn will follow overall economic conditions, as has been true for broad improvements in health and prosperity worldwide in the last two centuries. Some large foundations, such as Rockefeller and Gates, have explicitly followed a strategy of funding improvements in health to lead to overall economic development, but there is still much debate about whether that has worked.[12]

Despite these broad considerations, this history of blood transfusion in Africa has shown that it has received more attention and resources than other measures, such as basic public health infrastructure, that might have cost effectively saved more lives in the long run. Among the organizations continuing to assist those in transfusion services to gain attention and possibly additional resources to make the procedure more accessible to patients is the blood safety program of WHO, which has called for all countries to have a national blood policy. Another international organization that helps focus attention on these needs is the African Society for Blood Transfusion, created in 1997. Along with complementary organizations such as the Réseau d'Afrique Francophone de Transfusion, these also continue the tradition of meetings of African transfusion practitioners begun in the 1970s to exchange information and practices.

Beyond the technical and financial matters, perhaps the most interesting feature of transfusion will be how African countries organize the collection and donation of blood. The International Red Cross and WHO's blood safety committee, have consistently pushed for universal voluntary blood donation as the worldwide standard, which is justified on grounds of safety and efficiency.[13] Although this practice has scientific research to support it, the concept grows out the European historical experience with transfusion, which was very different from the history in Africa. This study has shown some of the complexities that will have to be taken into account as future policies and practices are developed.

NOTES

Abbreviations

AA	Archives africaines, Ministère des Affaires étrangères et de commerce extérieur, Brussels
ANS	Archives nationales du Sénégal
AOF	l'Afrique Occidentale Française
BRC Mechelen	Belgian Red Cross, Mechelen
BRC London	British Red Cross Museum and Archives, London
CNTS Dakar	Centre national de transfusion sanguine, Dakar
CRBC Brussels	Croix-Rouge du Congo, Section de la Croix-Rouge de Belgique (Congo Red Cross branch of Belgian Red Cross), Archives of the Belgian Red Cross, Belgian State Archives, Brussels
FRCBS	Finnish Red Cross Blood Service, Helsinki
IFRC	International Federation of Red Cross and Red Crescent Societies, Geneva
IMTSSA	Institut de médecine tropicale du service de santé des armées, Marseille
KNA	Kenya National Archives, Nairobi
KNH	Kenyatta National Hospital
KNBTS	Kenya National Blood Transfusion Service, Nairobi
MICR	Musée international de la Croix-Rouge, Geneva
ORL	Oxford Rhodes Library, Oxford
SRC	Swiss Red Cross, Bern
UBTS	Uganda Blood Transfusion Service, Kampala

Introduction

1. Paul M. Sharp and Beatrice H. Hahn, "Origins of HIV and the AIDS Pandemic," *Cold Spring Harbor Perspectives in Medicine* 1, no. 1 (September 2011): 1–22; 1:a006841. See also Tamara Giles-Vernick et al., "The Emergence of HIV/AIDS: An Historical Review," *Journal of African History* 54, no.1 (2013): 1–20.

2. C. Apetrei et al., "Molecular Epidemiology of Simian Immunodeficiency Virus SIVsm in U.S. Primate Centers Unravels the Origin of SIVmac and SIVstm," *Journal of Virology* 79, no. 14 (July 2005): 8991–9005.

3. Giles-Vernick et al. examine this question from a historical perspective. Giles-Vernick et al., "Emergence of HIV/AIDS." For papers at a recent international conference on the subject, see Guillaume Lachenal et al., "Simian Viruses and Emerging Diseases in Human Beings," *Lancet* 376, no. 9756 (December 4, 2010): 1901–2.

4. A. J. Nahmias et al., "Evidence for Human Infection with an HTLV III/LAV-like Virus in Central Africa, 1959," *Lancet* 1, no. 8492 (May 31, 1986): 1279–80; T. Zhu et al., "An African HIV-1 Sequence from 1959 and Implications for the Origin of the Epidemic," *Nature* 391, no. 6667 (February 5, 1998): 594–97.

5. Michael Worobey et al., "Direct Evidence of Extensive Diversity of HIV-1 in Kinshasa by 1960," *Nature* 455, no. 7213 (October 2, 2008): 661–64; Joel O. Wertheim and Michael Worobey, "Dating the Age of the SIV Lineages That Gave Rise to HIV-1 and HIV-2," *PLoS Computational Biology* 5, no. 5 (2009), e1000377; Philippe Lemey et al., "Tracing the Origin and History of the HIV-2 Epidemic," *Proceedings of the National Academy of Sciences of the United States of America* 100, no. 11 (May 2003): 6588–92.

6. Preston A. Marx, Cristian Apetrei, and Ernest Drucker, "AIDS as a Zoonosis? Confusion over the Origin of the Virus and the Origin of the Epidemics," *Journal of Medical Primatology* 33, no. 5–6 (October 2004): 220–26; Denis M. Tebit and Eric J. Arts, "Tracking a Century of Global Expansion and Evolution of HIV to Drive Understanding and to Combat Disease," *Lancet Infectious Diseases* 11, no. 1 (January 2011): 45–56.

7. Amit Chitnis, Diana Rawls, and Jim Moore, "Origin of HIV Type 1 in Colonial French Equatorial Africa?" *AIDS Research and Human Retroviruses* 16, no. 1 (January 2000): 5–8; Jim Moore, "The Puzzling Origins of AIDS," *American Scientist* 92, no. 6 (2004): 540–47.

8. Jacques Pepin, *The Origins of AIDS* (Cambridge: Cambridge University Press, 2011); Craig Timberg and Daniel Halperin, *Tinderbox: How the West Sparked the AIDS Epidemic and How the World Can Finally Overcome It* (New York: Penguin, 2012).

9. Giles-Vernick et al., "Emergence of HIV/AIDS."

10. Preston A. Marx, Phillip G. Alcabes, and Ernest Drucker, "Serial Human Passage of Simian Immunodeficiency Virus by Unsterile Injections and the Emergence of Epidemic Human Immunodeficiency Virus in Africa," *Philosophical Transactions of the Royal Society of London, B: Biological Sciences* 356, no. 1410 (2001): 911–20.

11. G. A. Lindboom, "The Story of a Blood Transfusion to a Pope," *Journal of the History of Medicine and Allied Sciences* 9, no. 4 (1954): 455–59 ; N. S. R. Maluf, "History of Blood Transfusion: The Use of Blood from Antiquity through the Eighteenth Century," *Journal of the History of Medicine and Allied Sciences* 9, no. 1 (1954): 59–107; "Injecting Blood Royal, or Phlebotomy at St. Cloud" (London: W. Holland, June 1804), *Images from the History of Medicine (NLM),* accessed April 24, 2012, http://ihm.nlm.nih.gov/images/A21765.

12. George Crile, *George Crile: An Autobiography,* ed. Grace Crile, 2 vols. (Philadelphia: Lippincott, 1947), 2:166.

13. Geoffrey Keynes, *Blood Transfusion* (London: Henry Frowde and Hodder and Stoughton, 1922).

14. See, for example, Cyrus C. Sturgis, "The History of Blood Transfusion," *Bulletin of the Medical Library Association* 30, no. 2 (January 1942): 105–12.

15. Susan E. Lederer, *Flesh and Blood: Organ Transplantation and Blood Transfusion in Twentieth-Century America* (New York: Oxford University Press, 2008).

16. "Editorial: Blood Transfusion Services," *South African Medical Journal* 49 (March 15, 1975): 379. These striking statistics indicating essentially a "white blood" supply were the result of a compromise policy to allay fears that whites would receive blood from black Africans, while at the same time permitting black Africans access to blood for transfusion. In the South African Blood Transfusion Service, which serves Johannesburg, whites donated 97 percent of blood, but 90 percent of recipients were nonwhite. There is surprisingly little scholarship on this history, leaving mainly memoirs and occasional commemorative pieces. See, for example, J. H. S. Gear, "Blood Transfusion in Johannesburg, Pre-South African Blood Transfusion Service," *Adler Museum Bulletin* 14, no. 1 (June 1988): 3–6.

17. For a preliminary report, see William H. Schneider and Ernest Drucker, "Blood Transfusions in the Early Years of AIDS in Sub-Saharan Africa," *American Journal of Public Health* 96, no. 6 (June 2006): 984–94.

18. Zaire was the name of the country from 1971 to 1997. After 1997 it became Democratic Republic of Congo, also its name from 1964 to 1971. From 1960 to 1964 it was called Republic of Congo.

Chapter 1: Transfusion before World War II

1. See, for example, Jean-Paul Roux, *Le sang: Mythes, symboles et réalités* (Paris: Fayard, 1988); and Thomas C. T. Buckley and Alma Gottlieb, *Blood Magic: The Anthropology of Menstruation* (Berkeley: University of California Press, 1988).

2. Mirko Grmek, *History of AIDS: Emergence and Origin of a Modern Pandemic,* trans. Russell C. Maulitz and Jacalyn Duffin (1989; repr., Princeton: Princeton University Press, 1990), 161. A more recent history is John Iliffe, *The African AIDS Epidemic: A History* (Athens: Ohio University Press, 2006).

3. J. Burke, "Historique de la lutte contre la maladie du sommeil au Congo," *Annales de la Société belge de médecine tropicale* 51, no. 4 (1971): 468.

4. John Iliffe, *East African Doctors: A History of the Modern Profession* (Cambridge: Cambridge University Press, 1998), 34–38. Although the mobilization was a health disaster, Iliffe points out it created "medical opportunities," especially for African subordinates. For a recent history, see Edward Paice, *World War I: The African Front* (New York: Pegasus, 2010).

5. Émile Lejeune, "Transfusion sanguine après hémoglobinurie grave," *Annales de la Société belge de médecine tropicale* 1 (1921): 299–300.

6. William H. Schneider, "Blood Transfusion in Peace and War, 1900–1918," *Social History of Medicine* 10, no. 1 (April 1997): 105–26. The first blood transfusion was reported in China in 1918 as well. See A. R. Kilgore and J. H. Liu, "Isoagglutination Tests of Chinese Bloods for Transfusion Compatibility," *China Medical Journal* 3 (1918): 21–25.

7. J. André, J. Burke, J. Vuylsteke, and H. van Balen, "Evolution of Health Services," in *Health in Central Africa since 1885: Past, Present and Future,* ed. P. G. Janssens, M. Kivits, and J. Vuylsteke (Brussels: King Baudouin Foundation, 1997), 71–74.

8. There has been a dramatic reassessment in the last two decades of the expansion of Western medicine in the nineteenth and twentieth centuries. Two recent overviews that serve as starting points are Ryan Johnson, "Historiography of Medicine in British Colonial Africa," *Global South* 6, no. 3 (2010): 20–28; and Deborah Joy Neill, *Networks in Tropical Medicine: Internationalism, Colonialism, and the Rise of a Medical Specialty, 1890–1930* (Stanford: Stanford University Press, 2012).

9. Anna Crozier, *Practising Colonial Medicine: The Colonial Medical Service in British East Africa* (New York: I. B. Tauris, 2007). For other studies of hospitals, see Mark Harrison, Margaret Jones, and Helen Sweet, eds., *From Western Medicine to Global Medicine: The Hospital beyond the West* (New Delhi: Orient Blackswan, 2009).

10. On the new more rapid testing technique developed at the Rockefeller Institute for Medical Research in 1915, see Peyton Rous and J. R. Turner, "A Rapid and Simple Method of Testing Donors for Transfusion," *Journal of the American Medical Association* 64, no. 24 (1915): 1980–82. On the wave of blood type testing around the world, see William H. Schneider, "The History of Research on Blood Group Genetics: Initial Discovery and Diffusion," *History and Philosophy of the Life Sciences,* 18, no. 3 (1996): 273–303.

11. The Germans elevated this practice to a procedure with the name *biologische Vorprobe.* See, for example, F. Oehlecker, "Erfahrungen aus 170 direkten Bluttransfusionen von Vene zu Vene," *Archiv für klinische Chirurgie,* 116 (1921): 714–15.

12. Lejeune, "Transfusion sanguine," 300. The Dieulafoy syringe was invented in 1869 to remove fluid from the pleural cavity.

13. André et al., "Evolution of Health Services," 118; Jean-Luc Vellut, "European Medicine in the Congo State (1885–1908)," in Janssens, Kivits, and Vuylsteke, *Health in Central Africa,* 72; G. Trolli, "Le service médical," in *L'essor économique belge: Expansion coloniale,* ed. Fernand Passelecq, 2 vols. (Brussels: Desmet-Verteneuil, 1932), 1:161–69. On Lejeune, see P. G. Janssens, "Eugène Jamot et Émile Lejeune: Pages d'histoire," *Annales de la Société belge de médecine tropicale* 75, no. 1 (March 1995): 10–11. On Trolli, see Jérôme Rodhain, "Giovanni Trolli: (24 juin 1876–8 février 1942)," *Bulletin l'Institut royal colonial belge* 17 (1946): 169–74; and for an organizational history, see Albert Dubois and Albert Duren, "Soixante ans d'organisation médicale au Congo belge," *Annales de la Société belge de médecine tropicale* 27 (1947): 1–36.

14. See, for example, Tanganyika Territory, *Annual Report of the Medical Department,* 1947, 28; Sierra Leone, *Annual Report, Medical and Health Services,* 1949, 17; 1950, 14.

15. Meghan Vaughan, "Health and Hegemony: Representation of Disease and the Creation of the Colonial Subject in Nyasaland," in *Contesting Colonial Hegemony: State and Society in Africa and India* , ed. Dagmar Engels and Shula Marks (London: British Academic Press, 1994), 185–86.

16. Meghan Vaughan, *Curing Their Ills: Colonial Power and African Illness* (Stanford: Stanford University Press, 1991), 58.

17. "(84) Barrett, Hutton, Philip W.," Medicine and Public Health in British Tropical Africa Documents, Oxford Rhodes Library (hereafter cited as ORL), Mss Afr s1872.

18. As quoted in George Crile, *George Crile: An Autobiography,* ed. Grace Crile, 2 vols. (Philadelphia: Lippincott, 1947), 2:166.

19. Joseph Lambillon and N. Denisoff, "Étude de l'organisation d'un service de transfusions sanguines dans un centre hospitalier d'Afrique," *Annales de la Société belge de médecine tropicale* 20 (1940): 279–85; D. Spedener, "Le traitement des pneumonies des noirs par transfusion de sang des convalescents," *Bulletin médical du Katanga* 1 (1924): 234–38.

20. K. S. Dewhurst, "Observations on East African Blood Donors," *East African Medical Journal* 22 (1945): 276–78.

21. Gaston Ouary, "Compte rendu de la Mission effectuée par le Médecin Commandant Ouary au Centre de Transfusion d'Alger (15 juillet–15 octobre 1944)," series H, 1H1 (1), Archives nationales du Sénégal, Dakar (hereafter cited as ANS), 20.

22. Hubert Carey Trowell, *Non-Infective Disease in Africa* (London: Edward Arnold, 1960), 426n1.

23. Schneider, "Blood Transfusion in Peace and War," 105–26. For broader developments in the United States, see Susan E. Lederer, *Flesh and Blood: Organ Transplantation and Blood Transfusion in Twentieth-Century America* (New York: Oxford University Press, 2008).

24. Kim Pelis, "Taking Credit: The Canadian Army Medical Corps and the British Conversion to Blood Transfusion in WWI," *Journal of the History of Medicine and Allied Sciences* 56, no. 3 (2001): 238–77; Schneider, "Blood Transfusion in Peace and War."

25. William H. Schneider, "Blood Transfusion between the Wars," *Journal of the History of Medicine and Allied Sciences* 58, no. 2 (April 2003): 197–207.

26. Gunnar Alsted, "Red Cross Participation in Blood Transfusion," in *Fifth International Congress of Blood Transfusion, Paris 1954* (Paris: Édition Septembre, 1955), 185–88; J. Spaander, "Le rôle de la Croix-Rouge néerlandaise dans le domaine de la transfusion sanguine," *Revue d'hématologie* 5, nos. 3–4 (1950): 500; M. Adant, "L'organisation de la transfusion sanguine en Belgique," in *Fifth International Congress of Blood Transfusion, Paris 1954* (Paris: Édition Septembre, 1955), 999.

27. American Association of Blood Banks, *The 2005 Nationwide Blood Collection and Utilization Survey Report* (Bethesda, MD: AABB, 2005), accessed March 13, 2013, http://www.aabb.org/programs/biovigilence/nbcus/Documents/05nbcusrpt.pdf.

28. For France, see Établissement français du sang, "Donner son sang en France, quatrième edition, Juillet 2007" (Paris: Établissement français du sang), accessed August 26, 2011, http://www.cerphi.org/wp-content/uploads/2011/05/Don-sang-2007.pdf. For Britain, see H. Dodsworth and H. H. Gunson, "Fifty Years of Blood Transfusion," *Transfusion Medicine* 6, suppl. 1 (1996): 1–88.

29. Tibor Greenwalt, *History of the International Society of Blood Transfusion, 1935—1995* (Groningen: Stichting Transfusion Today Foundation, 2000); Juhani Leikola and W. G. van Aken, "The Story of *Vox sanguinis,*" *Vox sanguinis* 100, no.1 (2011): 2–9.

30. With no overview assessment of the whole continent, one must rely of surveys by colonial power or region. For the Congo, see Nancy Rose Hunt, *A Colonial Lexicon of Birth Ritual, Medicalization, and Mobility in the Congo* (Durham, NC: Duke University Press, 1999); J. André et al., "Evolution of Health Services," 89–158. On French colonial medicine, see Jean-Paul Bado, *Les conquêtes de la médecine moderne en Afrique* (Paris: Karthala, 2006); Myron Echenberg, *Black Death, White Medicine: Bubonic Plague and the Politics of Public Health in Colonial Senegal, 1914–1945* (Portsmouth, NH: Heinemann,

2002) Two book-length studies of British Africa are Meghan Vaughan, *Curing Their Ills*, and Ralph Schram, *A History of the Nigerian Health Services* (Ibadan: Ibadan University Press, 1971).

31. Rodhain, "Giovanni Trolli"; Dubois and Duren, "Soixante ans"; André et al., "Evolution of Health Services," 89–158. On French colonial medicine, see Jean-Paul Bado, *Les conquêtes de la médecine moderne en Afrique* (Paris: Karthala, 2006); Myron Echenberg, *Black Death, White Medicine: Bubonic Plague and the Politics of Public Health in Colonial Senegal, 1914–1945* (Portsmouth, NH: Heinemann, 2002) 118; Maryinez Lyons, "Public Health in Colonial Africa: The Belgian Congo," in *The History of Public Health and the Modern State*, ed. Dorothy Porter (Amsterdam: Editions Rodopi, 1994), 371.

32. Henri Brunschwig, *Noirs et blancs dans l'Afrique noire française, ou, Comment le colonisé devient colonisateur, 1870–1914* (Paris: Flammarion, 1983), 73.

33. On dressers, see Crozier, *Practising Colonial Medicine*, 83–84. Raymond Barrett, a British doctor who served in Uganda and Tanganyika from 1928 to 1955, claimed that the subassistant surgeons, who did a three-year training course in India, "played an indispensable role" in all district hospitals. See "(9) Barrett, Raymond E.," Medicine and Public Health in British Tropical Africa Documents, ORL, Mss Afr s1872. See also John A. Carman, *A Medical History of the Colony and Protectorate of Kenya: A Personal Memoir* (London: Rex Collings, 1976), 6–9; F. M. Mburu, "Socio-Political Imperatives in the History of Health Development in Kenya," *Social Science and Medicine, A: Medical Sociology* 15, no. 5 (1981): 524. On the Dakar medical school, see Maurice Blanchard, "L'École de médecine de l'Afrique Occidentale Française de sa fondation à l'année 1934," *Annales de médecine et de pharmacie coloniale* 33 (1934): 90–111; "L'école de médecine indigène de l'Afrique occidentale française," *Bulletin d'informations et de renseignements* (AOF) 199 (August 15, 1938): 303–6.

34. Blanchard, "École de médecine," 94–95.

35. Claude Chippaux, "Le service de santé des troupes de marine," *Médecine tropicale* 40, no. 6 (1980): 616.

36. Schneider, "Blood Transfusion between the Wars."

37. Schneider, "Blood Group Genetics."

38. James Hunter Harvey Pirie, "Blood Testing Preliminary to Transfusion, with a Note on the Group Distribution among S.A. Natives," *Medical Journal of South Africa* 16 (January 1921): 109.

39. On the medical facilities of the Union minière du Haut Katanga, see L. Mottoulle, "L'organisation du Service médical et la situation sanitaire générale à l'Union minière du Haut-Katanga, fin 1929," *Annales de la Société belge de médecine tropicale* 11 (1931): 253–55. On Belgian investment in health in the Congo more generally, see Hunt, *Colonial Lexicon*.

40. See André et al., "Evolution of Health Services."

41. The main hospital for Africans in Léopoldville (later Kinshasa) went by different names at different times, such as Hôpital des noirs, later Hôpital général, Mama Yemo. For convenience, Hôpital des congolais will be used during the colonial period.

42. A. Lodewyck, "Note sur la transfusion sanguine chez les nourrissons et les enfants," *Recueil de travaux de sciences médicales au Congo belge* 2 (1944): 157–61.

43. "Rapport des Services Médicaux de la Province de Léopoldville Année 1956," p. 256, RA/MED 18, Archives africaines, Ministère des affaires étrangères et de commerce extérieur, Brussels, hereafter AA.

44. A rough correlation between ABO tests and transfusions is about three tests per transfusion: one for the patient and two tests to get a match from a donor.

45. Lambillon and Denisoff, "Service de transfusions sanguines," 279–85.

46. C.-S. Ronsée, "Sur les anémies malariennes des enfants et les transfusions sanguines," *Comptes rendus du congrès scientifique, Elisabethville, 13–19 août 1950,* vol. 5, *Travaux de la Commission de médecine humaine et vétérinaire* (Brussels: Comité spécial du Katanga, 1950): 96–107.

47. Spedener, "Traitement des pneumonies"; Germond, "Statistiques des cas de pneumonie traités par transfusion de sang de convalescents," *Bulletin médical du Katanga* 1 (1924): 243; R. van Nitsen, "La pneumonie chez le noir: Essai de traitement," *Bulletin médical du Katanga* 1 (1924): 239–42.

48. Van Nitsen, "La pneumonie," 242.

49. George Valcke, "Note," *Annales de la Société belge de médecine tropicale* 14 (1934): 432–33.

50. A. Duboccage, "Notes cliniques du service gynécologique de la Fomulac à Kisantu," *Annales de la Société belge de médecine tropicale* 14 (1934): 421–33.

51. L. Kok, "Les transfusions de sang en milieu indigène" (medical thesis, Prince Leopold Institute of Tropical Medicine, Antwerp, 1950), 1.

52. Ibid., 2–5.

53. For a prosopography of colonial medical officers, see Crozier, *Practising Colonial Medicine.* An excellent bibliography on the introduction of Western medicine in East Africa is also in Iliffe, *East African Doctors.*

54. As cited in Osaak Olumwullah, *Dis-ease in the Colonial State: Medicine, Society, and Social Change among the AbaNyole of Western Kenya* (Westport, CT: Greenwood, 2002), 182.

55. Crozier, *Practising Colonial Medicine,* 40.

56. Schneider, "Blood Transfusion between the Wars," 197–207.

57. *British Empire Red Cross Conference 1930* (London: British Red Cross Society, 1930), 31. For more on Oliver, see G. W. Bird, "Percy Lane Oliver, OBE (1878–1944): Founder of the First Voluntary Blood Donor Panel," *Transfusion Medicine* 2, no. 2 (June 1992): 159–60.

58. *British Empire Red Cross Conference 1930,* 31.

59. British Red Cross Society, *Annual Report,* 1932, 40.

60. As cited in Mwelwa C. Musambachime, "The Impact of Rumor: The Case of the Banyama (Vampire Men) Scare in Northern Rhodesia, 1930–1964," *International Journal of African Historical Studies* 21, no. 2 (1988): 208. For more, see chapter 5.

61. British Red Cross Society, *Annual Report,* 1939, 92, 95; Kenya, Colony and Protectorate, *Medical Research Laboratory Annual Report,* 1933–37.

62. "Rapport annuel, Hôpital Central Africain Année 1949," Institut de médecine tropicale du service de santé des armées Marseille (hereafter IMTSSA), box 32; "Rapport annuel Hôpital Principal Année 1949," IMTSSA, box 33. For a few years le Dantec included annual reports of the reanimation–transfusion service, see 1950–52.

On French Congo, see "Inspection générale du Service de Santé, AEF Colonie du Moyen-Congo, Rapport Annuel 1933," 111; 1934, 124, IMTSSA, box 117.

63. Spedener, "Traitement des pneumonies"; Germond, "Statistiques"; Uganda Protectorate, *Annual Medical and Sanitary Report* 1931, 49; Kenya, Colony and Protectorate, *Medical Research Laboratory Annual Report,* 1933; Tanganyika Territory, *Annual Medical and Sanitary Report,* 1932, 67; "Inspection Générale du Service de Santé. AEF Colonie du Moyen-Congo Rapport Annuel 1933," 111; 1934, 124, IMTSSA, box 117; R. Ghose, "History of Blood Transfusion in Ethiopia," *Ethiopian Medical Journal* 31, no. 4 (April 1963): 208; Gold Coast Colony, *Departmental Reports, 1935–36,* 44; Sierra Leone, *Annual Report of the Medical and Sanitary Department,* 1936, 51; British Red Cross Society, *Annual Report,* 1939, 92, 95; l'Afrique Occidentale Française (hereafter AOF), Service de santé, *Rapport annuel,* 1940, 79; AOF, "Inspection générale des services sanitaires et médicaux, Rapport annuel 1941," IMTSSA, box 4, 115.

64. Uganda Protectorate, *Annual Medical and Sanitary Report,* 1931, 49; Tanganyika Territory, *Annual Medical and Sanitary Report,* 1932, 67; Gold Coast Colony, *Departmental Reports,* 1935–36, 44.

65. AOF, "Inspection générale des services sanitaires et médicaux, Rapport annuel 1941," 115.

66. Service médical Union minière du Haut Katanga, *Rapport annuel,* 1929, 1932, 1939; Service de l'hygiène et laboratories bacteriologique, Léopoldville, *Rapport annuel,* 1937.

Chapter 2: Blood Transfusion from 1945 to Independence

1. "Documentation concernant la Santé publique en AOF," 1953, box 84, table 3 and table 3bis, 17–19, Institut de médecine tropicale du service de santé des armées, Marseille, France, hereafter IMTSSA. For background, see also Martin-René Atangana, *French Investment in Colonial Cameroon: The FIDES Era (1946–1957)* (New York: Peter Lang, 2009); J. André, J. Burke, J. Vuylsteke, and H. Van Balen, "Evolution of Health Services," in *Health in Central Africa since 1885: Past, Present and Future,* ed. P. G. Janssens, M. Kivits, and J. Vuylsteke (Brussels: King Baudouin Foundation, 1997), 125–28.

2. Adetokunbo A. Lucas, "What We Inherited: An Evaluation of What Was Left Behind at Independence and Its Effects on Health and Medicine Subsequently," in *Health in Tropical Africa during the Colonial Period,* ed. E. E. Sabben-Clare, David J. Bradley, and Kenneth Kirkwood (Oxford: Clarendon Press, 1980), 239–248; André Prost, *Services de santé en pays africain: Leur place dans des structures socio-économiques en voie de développement* (Paris: Masson, 1970); Ralph Schram, *A History of the Nigerian Health Services* (Ibadan: Ibadan University Press, 1971).

3. A large-scale blood bank was established in Bombay in conjunction with the Haffkine Institute. See British Red Cross Society, *Annual Report,* 1942–45. Unlike in Australia and Canada, however, the Indian service was disbanded after the war.

4. Gaston Ouary, "Compte rendu de la mission effectué par le Médecin Commandant Ouary au Centre de Transfusion d'Alger (15 juillet–15 octobre 1944)," 18, series H, 1H1 (1), Archives nationales du Sénégal, Dakar, hereafter ANS.

5. Ibid., 4–5.

6. This has continued to be a heated debate. For an unusually balanced view, see Lucas, "What We Inherited"; for an example of the contemporary critique, see

Anne-Emanuelle Birn, "Gates's Grandest Challenge: Transcending Technology as Public Health Ideology," *Lancet* 366, no. 9484 (August 6, 2005): 514–19.

7. John Iliffe, *East African Doctors: A History of the Modern Profession* (Cambridge: Cambridge University Press, 1998), 136–37, 174–75.

8. J. André et al., "Evolution of Health Services," 140.

9. For the American Red Cross, see Foster Rhea Dulles, *The American Red Cross: A History* (New York: Harper, 1950); for Britain and the Commonwealth, see *British Red Cross Society Quarterly Review* for 1939–45.

10. On transfusion in France, see Sophie Chauveau, "De la transfusion à l'industrie: Une histoire des produits sanguins en France (1950–fin des années 1970)," *Entreprises et histoire* 2, no. 36 (October 2004): 103–19; and her "Du don à l'industrie: La transfusion sanguine en France depuis les années 1940," *Terrain: Revue d'ethnologie de l'Europe,* no. 56 (March 2011): 74–89. Ironically, the British followed suit after the Second World War when the National Health Service took charge of transfusion. See William H. Schneider, "Blood Transfusion between the Wars," *Journal of the History of Medicine and Allied Sciences* 58, no. 2 (April 2003): 187–224.

11. Schneider, "Blood Transfusion between the Wars," 206–7.

12. On Lambillon, see Dr. A. Duren, "Note pour le service du personnel d'Afrique," May 31, 1945, folder "Fomulac personnel médical," in Hygiene 4452/808 Fomulac, 51AI, Archives africaines, Ministère des affaires étrangères et de commerce extérieur, Brussels, hereafter AA. On Ouary in Gabon, see Territoire du Gabon, "Rapport médicale 1951," 3, box 128, IMTSSA; on Madagascar, see "Correspondences (Jan.–Dec. 1955)," box 280, Centres de transfusion, IMTSSA.

13. Myriam Malengreau, pers. comm. to author, describing phone conversations with Legrand, February 19, 2006. Legrand served at Katana Hospital in Kivu from 1952 to 1960.

14. C.-S. Ronsée, "Anémies malariennes des enfants et transfusions sanguines, avec observations sur les groupes sanguins des Bakongo," *Mémoires Institut royal colonial belge, Section des sciences naturelles et médicales* 20, no. 2 (1952): 1–64.

15. On Ibadan, see Alan F. Fleming, "Memoir on University College Hospital, Ibadan, and the University of Ibadan, December 1962–July 1966," October 3, 1983, in Medicine and Public Health in British Tropical Africa Documents, Mss Afr s1872, no. 52, Oxford Rhodes Library, Oxford, hereafter ORL. "Una Maclean, medical officer, Blood Transfusion Lab, University College Hospital, Ibadan, 1957–59," May 10, 1983, Mss Afr s1872, no. 99, ORL. On blood transfusions at the new hospital at Lomé, Togo, built in the early 1950s with FIDES funds, see "Rapport annuel du gouvernement français à l'Assemblé général des Nations Unies sur l'administration du Togo, 1955," box 25, IMTSSA. On the new "salle de réanimation" installed at the Hôpital indigène in Dakar, see "Hôpital indigène de Dakar, Rapport annuel, 1949," 3, AOF MiOM/1827 (2G41–9), Archives d'outre-mer, Aix-en-Provence.

16. For an example of the French government's control of the amount to be paid to official blood donors in the colonies, see "Donneurs du sang, lettres circulaires, 9513 et 9514," October 20, 1950, box 322, IMTSSA.

17. On reports of payments of 5 shillings per donor in Nyasaland, see Pat Jephson to Joan Whittington, February 25, 1961; Secretary General to Miss Whittington, March

9, 1961, Acc 0076/39(3), British Red Cross Museum and Archives, London, hereafter BRC London.

18. "Rapport Annuel, Institut Pasteur de Dakar," 1943, 76, Services des Archives de l'Institut Pasteur, Paris. See also Jacques Linhard, "Le centre fédéral de transfusion de l'AOF," *Médecine tropicale* 11, no. 6 (1951): 951–57; Etienne G. Ayité, G. Diebolt, and J. Linhard, "La transfusion sanguine au Sénégal," *Bulletin de la Société médicale d'Afrique noire de langue française* 18, no. 3 (1973): 289–92.

19. In addition to Ouary, "Compte rendu," see Edmond Benhamou's contribution to the following special issue, "Notes pour servir à l'histoire de la transfusion sanguine dans l'armée française de 1942 à 1945 à partir de l'Afrique du Nord," *Revue des Corps de santé des Armées Terre, Mer, Air* 7 (November 1966): 859–62.

20. "Rapport annuel, Hôpital Central Africain, partie médicale," 1949–51, box 32, IMTSSA; "Hôpital principal de Dakar, Rapport annuel," 1949, box 32, IMTSSA also mentions establishment of a reanimation-transfusion service, but no reports of activity were given.

21. Gaston Ouary, Yann Goez, and Jacques Linhard, *Notes sur la réanimation-transfusion* (Algiers: Service de santé des troupes coloniales, 1944).

22. Ibid.

23. "Avant-propos," ibid.

24. Ouary, "Compte rendu," 7.

25. Linhard, "Centre fédéral," 951–52.

26. Rapport annuel, Hôpital central africain, années 1949, 1950, 1951, box 32, IMTSSA.

27. Linhard, "Centre fédéral," 951–52. For an example of another French colony (Tahiti) creating a transfusion service at this same time, see J. Lhoiry, "L'utilisation de la transfusion sanguine outre-mer dans les centres chirurgicaux secondaires," *Médecine tropicale* 14, no. 5 (1954): 569–79.

28. Le Rouzic to Governor General, August 9, 1949, series H, 1H1 (1), ANS.

29. "Centre Fédéral de Transfusion Sanguine, Rapport annuel, 1954," box 84, IMTSSA. In addition to the published accounts by Linhard cited above, a brief history can also be found in "Les Centres de Transfusion Sanguine dans les territories de la France d'Outre-Mer et en Extrême Orient," box 280, IMTSSA.

30. Being the only such facility in the colonies, the Dakar service was an important source of plasma for troops fighting during the height of the French Indochina War, in the early 1950s. Former colonial medical officers report problems with keeping the equipment running and with contamination of the plasma that caused some hospitals to reject it in favor of whole blood. Whatever the reason, the shipments of dried plasma declined throughout the 1950s.

31. "Rapport annuel, Hôpital Central Africain," 1950–52, box 32, IMTSSA.

32. Inspection générale du service de santé, AEF Colonie du Moyen-Congo, Rapport annuel, 1933, 1934, box 117, IMTSSA.

33. Territoire du Gabon, "Rapport Médicale," 1950, 130, box 128, IMTSSA; "Rapport Médicale," 1951, box 269, IMTSSA; President, Fédération nationale des donneurs du sang de France et d'Outre-Mer to Médecin-Général Salaun, Directeur du Service de santé coloniale, "Circulaires," October 3, 1950, box 322, IMTSSA.

34. Territoire du Gabon, "Rapport médicale," 1951, 269, box 128, IMTSSA; "Centres de transfusion sanguine," August 19, 1954, box 280, IMTSSA.

35. Croix-Rouge togolaise de Lomé, "Rapport d'activité," 1954, AO917, Archives of International Federation of Red Cross and Red Crescent Societies, Geneva, hereafter IFRC.

36. Unclassified records, Centre national de transfusion sanguine, Dakar, hereafter CNTS Dakar.

37. Annual medical reports for each colony, 1955, 1956, IMTSSA.

38. "Assemblé Territoriale de la Côte d'Ivoire, projet de deliberation tendant à la creation en Côte d'Ivoire d'un Centre de Transfusion Sanguine," September 13, 1957, box 294, IMTSSA.

39. The "Projet de Budget Gestion, 67/68" lists 18 million francs CFA for "indemnités aux donneurs" in the annual budget, CNTS Dakar.

40. *British Empire Red Cross Conference 1930* (London: British Red Cross Society, 1930), 31.

41. In the United States, blood was segregated, not only locally but also by Red Cross policy during WWII and in the U.S. army until the early 1950s. See Schneider, "Blood Transfusion between the Wars," 221–22.

42. *Southern Rhodesia Red Cross Annual Report, 1939,* 7–8, Acc 0076/51 Southern Rhodesia, 1936–56, BRC London.

43. Based on published annual reports of colonial government medical departments, and Red Cross reports, 1947–62, BRC London.

44. Joan Whittington, "Report on Southern Rhodesia Central Council branch," June 9, 1948, Acc 0076/51, BRC London; *British Red Cross Society, Overseas Branches,* 1950, 42 BRC London. Because of racial tension, only whites were allowed to donate for transfusions in South Africa and, during the Second World War, in the United States.

45. "British Red Cross Society, Kenya (Central Council) Branch Report for the year ending 31 October 1948," Acc 0287/37 Kenya, BRC London; Kenya, Colony and Protectorate, "Medical Department Annual Report," 1947, 63 BRC London; ibid., 1948, 66.

46. M. G. MacVicar, "Northern Rhodesia Report No. 3," December 9, 1950, Acc 0076/38(1) Northern Rhodesia, 1940–61, BRC London. For Enugu's first blood bank in Nigeria, see "Nigerian Branch News Sheet, no. 3," April 1953, Acc 0076/36(1), box 7, BRC London; for Lagos and Ibadan, see ibid., no. 5, October 1953.

47. Joan Whittington "Report on the Nyasaland Local Branch" June 9, 1948, Acc 0287/46 Nyasaland; Whittington, "Report on visit to Tanganyika Territory," June 9, 1948, Acc 0287/60 Tanganyika; "Report for the Year 1948 from the Uganda Central Council branch," Acc 0287/63 Uganda; "Report to British Red Cross from Lusaka," October 25, 1949 Acc 0076/38(1); "Miss Borley's report," November 1950, Acc 0076/6(1); Gold Coast, "Summary Report for 1952," Acc 0287/33 Gold Coast; Nigerian Central Council, "Annual Report," 1952, 5, Acc 0076/36(1); Sierra Leone Branch Red Cross Society, "Annual Report," 1956, Acc 0076/48(2); M. D. N'Jie, "Red Cross Week, 9th—14th March, 1959," March 20, 1959, Acc 0076/21(2) Gambia, all in BRC London.

48. For background that focuses on recent history, see E. J. Watson-Williams and P. K. Kataaha, "Revival of the Ugandan Blood Transfusion System 1989: An Example of International Cooperation," *Transfusion Science* 11, no. 2 (1990): 179–84.

49. "Blood Transfusion Service: Sub-committee of BMA in Blood Transfusion, meeting held on 4th May 1948," Acc 0076/58(1) Uganda, 1949–1956, BRCS.

50. On Mengo Hospital, see Elizabeth Moody, "Blood Transfusion in Uganda," *British Red Cross Society Quarterly Review* 36 (1949): 107.

51. "Blood Transfusion Service: Sub-committee of BMA in Blood Transfusion, meeting held on 4th May 1948," Acc 0076/58(1) Uganda, 1949–1956, BRC London.

52. Ibid.; "Report for the Year 1948 from the Uganda Central Council branch," Acc0287/63 Uganda, BRC London.

53. Moody, "Blood Transfusion in Uganda," 109.

54. "Uganda Central Council Red Cross Branch Annual report 1949–50," Acc 0076/58(1) Uganda, 1949–56, BRC London.

55. Uganda Protectorate, *Annual Medical and Sanitary Report,* 1957, 39.

56. British Red Cross Society (Uganda Branch), Blood Transfusion Service, "Annual Report," 1956, 3, Acc 0076/58(2) Uganda, BRC London.

57. British Red Cross Uganda Branch, "Annual Report," 1959, 1965; "Uganda Blood Transfusion Annual Report," 1962, 1963, Acc 0076/58(2) Uganda, BRC London.

58. Moody, "Blood Transfusion in Uganda," 107.

59. "Speech by His Excellency the Governor at the annual general meeting of the Uganda Red Cross Society, Kampala, 24th April 1952," Acc 0076/58(1) Uganda, 1949–56, BRC London.

60. Maltby to Whittington, December 22, 1959, Acc 0076/58(3) Uganda, BRC London.

61. See correspondence and reports, 1957–66, Acc 0076/61(1) Zambia, BRC London.

62. Kenya, Administrative Reports, "Medical Department Annual Report," 1957, 8; ibid., 1959, 26.

63. "To all divisions, Kenya Branch British Red Cross Society, Blood Donor Service," April 1960, BY4/69, Kenya National Archives, Nairobi, hereafter KNA. The timing also followed reports of accidents in blood transfusion, plus ultimately the death of a European. See for example, W. E. Lawes, Consultant Anesthetist at King George VI Hospital to Chair, Surgical Committee, "Blood Bank Services," April 7, 1959, R26, KNA.

64. Lewis to Rigby, February 22, 1961, BY4/69, KNA.

65. Tanganyika Territory, *Annual Report of the Medical Department [Laboratory],* 1957, 11; ibid., 1958, 9.

66. For more on G. M. Edington (1916–81), who continued his career in Nigeria after leaving the Gold Coast at independence, see his obituary in the *London Times,* February 13, 1981, 16.

67. Gold Coast Branch, *Annual Report,* 1952, 1953, Acc 0287/33 Gold Coast, BRC London.

68. Gold Coast Colony, *Medical Research Institute Report,* 1954–57.

69. Antoine Duboccage, "La Fondation médicale de l'Université de Louvain au Congo," *Lovania* 3 (1944): 6–10; Joseph Lambillon, "Fomulac Katana: Deuxième centre

de la Fondation médicale de l'Université de Louvain au Congo," *Lovania* 5 (1944): 5–10; Paul Norbert Hennebert, "La médicine à Lovanium," accessed September 25, 2006, http://www.md.ucl.ac.be/histoire/livre/lovan.pdf.

70. Malengreau to Minister of Colonies, September 29, 1936, Hygiène 4452/808 Fomulac, AA.

71. Joseph Lambillon and N. Denisoff, "Étude de l'organisation d'un service de transfusions sanguines dans un centre hospitalier d'Afrique," *Annales de la Société belge de médecine tropicale* 20 (1940): 279–85.

72. On Lambillon's research about eclampsia, see Nancy Rose Hunt, "Normality and a 'Disease of Civilization': Eclampsia and Race in the Congo and the U.S. South," in *Discovering Normality in Health and the Reproductive Body,* proceedings of a workshop held at the Program of African Studies, Northwestern University, March 9–10, 2001, ed. Caroline H. Bledsoe (Evanston: Program on African Studies, 2002), 125–35.

73. A brief account was published by Albert Dubois, *La Croix Rouge au Congo, Classes des sciences naturelles et médicales,* new series 18–2 (Brussels: Académie royale des sciences d'outre-mer, 1969),: 1–61. In addition, see Alexandre Michiels, "L'oeuvre de la Croix-Rouge au Congo belge (1889–1960)" (Memoire de license en Histoire, Université de Bruxelles, 2000), copy deposited in archives of the Archives Croix-Rouge du Congo, Section de la Croix-Rouge de Belgique (Congo Red Cross branch of the Belgian Red Cross), Archives of the Belgian Red Cross, Belgian State Archives, Brussels, hereafter CRBC Brussels.

74. "Comptes Rendus du Congo," *Rapport annuel, Croix-Rouge du Congo,* 1952, 27, RK Congo, box 1, CRBC Brussels. The figures on major operations were 334 for the hospital at Medje, 322 at Pawa, and 170 at Irambi.

75. *Rapport annuel, Croix-Rouge du Congo,* 1940–45, 23, RK Congo, box 1, CRBC Brussels.

76. See *Rapport annuel, Croix-Rouge du Congo,* 1945–47, RK Congo, box 1, CRBC Brussels. On the concern with maternity and infant health, see Nancy R. Hunt, *A Colonial Lexicon of Birth Ritual, Medicalization, and Mobility in the Congo* (Durham, NC: Duke University Press, 1999). See also "Recapitulation [of CR Congo account balance]," June 26, 1945, RK Congo, box 4, CRBC Brussels.

77. There is very little biographical information about these two doctors. The archives of the Congo Red Cross contain a large personnel dossier on Claude Lambotte: CR Congo, box 23, personnel dossiers, CRBC Brussels. When his wife, Jeanne, died, in 1960, a brief obituary was published by Albert Dubois, former director of the Prince Leopold Institute of Tropical Medicine. Dubois, "Jeanne Legrand (1916–60)," *Annales de la Société belge de médecine tropicale* 40 (1960): 711. In the archives and publications her last name is sometimes given as her maiden name (Legrand) or as Lambotte-Legrand, as well as her husband's name. For the sake of consistency, I have used her married name, Lambotte.

78. "Rapport annuel, Hôpital Central Africain, année 1949," 52–55, box 32, IMTSSA.

79. "Maclean, Una (b. 1925), medical officer, Blood Transfusion Lab, University College Hospital, Ibadan, 1957–59," May 10, 1983, Mss Afr s1872, no. 99, ORL.

80. Province de Katanga, Service de l'hygiène, "Rapport annuel," 1947, 72, RA/MED 19, AA.

81. Province de Léopoldville, Service de l'hygiène, "Rapport annuel," 1947, 219, RA/MED 16, AA; Province de Katanga, Service de l'hygiène, "Rapport annuel," 1947, 124, RA/MED 19, AA.

82. *Rapport annuel, Croix-Rouge du Congo,* 1947–53, CRBC Brussels; *Voix du congolais,* January 1954, 54.

83. The Léopold-Est hospital had 575 beds. The next largest in the colony was the "Hôpital des indigènes" at Elizabethville, with 342 beds. Institut de Médecine Tropicale Princesse Astrid, Léopoldville, "Rapport annuel," 1954, A11 Hygiène, 4470. A 1949 plan first envisioned a fifteen hundred–bed hospital for Léopoldville, but it was scaled back to one thousand beds. *Courrier d'Afrique* (January 13, 1958), A11 Hygiène 4472, Belgium, AA.

84. L. van Hoof to Malengreau, April 26, 1945, 813 Fomulac au Kivu, AA.

85. Edouard Dronsart, "Voyage au Congo, Stanleyville—Pawa—Elisabethville—Leopoldville," 1951, RK Congo, foto box 32, CRBC Brussels.

86. Joseph Lambillon, "Contribution à l'étude du problème obstétrical chez l'autochtone du Congo belge," *Annales de la Société belge de médecine tropicale* 30, no. 5 (1950): 987–1123. See also Hunt, "Normality."

87. Croix-Rouge du Congo, "Rapport d'activité, 1948," 28, RK Congo, box 1, CRBC Brussels. For pictures of Africans lining up to register and file card records, see RK Congo, foto box 30, CRBC Brussels.

88. Croix-Rouge du Congo, "Rapport d'activité," 1957, 37; "Rapport d'activité," 1927–59, both in RK Congo, box 1, CRBC Brussels.

89. Dronsart to Lambotte, August 10, 1951; Lambert to Dronsart, September 14, 1951; Dronsart to Lambotte, September 19, 1951, all in RK Congo, box 2, CRBC Brussels.

90. Croix-Rouge du Congo, "Rapport d'activité," 1953, 39, RK Congo, box 1, CRBC Brussels.

91. "Process-verbaux, Croix-Rouge du Congo Réunion du Comité Exécutif, 21 November 1953," *Rapport annuel, Croix-Rouge du Congo,* 1954, 47, RK Congo, box 3, CRBC Brussels.

92. *L'avenir colonial,* March 7, 1955, press clippings, RK Congo, box 34, CRBC Brussels.

93. "Process-verbaux, Croix-Rouge du Congo Réunion du Comité Exécutif du 20 décembre 1951," *Rapport annuel, Croix-Rouge du Congo,* 1951, 2, RK Congo, box 3, CRBC Brussels. A request from Stanleyville for a similar organization was rejected.

94. For press coverage, see *L'avenir colonial,* April 7–8, 1951, March 7, 1955; *Voix du congolais,* October 1955, 551–56, RK Congo, box 34, CRBC Brussels. The latter has a picture of Jeanne Lambotte and the van.

95. See "La Croix-Rouge du Congo organise à Léopoldville, un service de transfusion sanguine," *Voix du congolais,* June 1955, 509–12, RK Congo, box 34, CRBC Brussels.

96. *Rapport annuel, Croix-Rouge du Congo,* 1954, 48–49, RK Congo, box 1, CRBC Brussels.

97. "Process-verbaux, Croix-Rouge du Congo Réunion du Comité Exécutif, 15 February 1956," *Rapport annuel, Croix-Rouge du Congo,* 1956, RK Congo, box 3, CRBC Brussels.

98. *Rapport annuel, Croix-Rouge du Congo,* 1957, 40–41, RK Congo, box 1, CRBC Brussels; *L'Afrique et le monde,* March 21, 1957, 1, RK Congo, box 34, CRBC Brussels. See also Kivits to Gouverneur Général, January 3, 1957, recommending a transfusion service; Gouverneur Général to Kivits, January 27, 1957, approving his recommendation, RA/MED 85, folder 1256, Croix-Rouge du Congo—Centre de transfusion sanguine, AA.

99. See report of Service de Transfusions Sanguine à Léopoldville, July 1957, RA/MED 85, folder 1256, Croix-Rouge du Congo—Centre de transfusion sanguine, AA.

100. Lambillon to Kadaner, November 5, 1957, folder 1256, "Croix-Rouge du Congo—divers," Hygiène 4519, AA.

101. Institut de Médecine Tropicale Princesse Astrid, Léopoldville, *Rapport annuel,* 1954, A11 Hygiène 4470, AA; "Rapport des Services Médicaux de la Province de Léopoldville," 1956, RA/MED 1, AA.

102. "Process-verbaux, Croix-Rouge du Congo, Réunion du Comité Exécutif, 15 February 1956," *Rapport annuel, Croix-Rouge du Congo,* 1956, RK Congo, box 3, CRBC Brussels.

103. Beginning November 19, 1957, and continuing for over a year, the monthly meetings of the committee described the heated discussion and efforts to reach a settlement.

104. This evidence can be found in numerous reports cited above, both published as well as in the CRBC Brussels and AA archives.

105. "Rapport annuel, Institut Pasteur de Cameroun," 1960, box 39, IPO-RAP, Archives de l'Institut Pasteur, Paris; "Territoire du Tchad," Service de santé, 1960, box 123, IMTSSA; Croix-Rouge du Congo, Section de la Croix-Rouge de Belgique, rapport, 1959, CRBC Brussels; Congo, Brazzaville, Service national de la statistique, Annuaire statistique, 1960, IMTSSA; Dahomey, Ministère de la santé publique, *Rapport médical sur l'activité du service de santé de la république, 1962*; Gold Coast Colony, *Medical Research Institute Report,* 1957; Baba Sy, "Fonctionnement de la banque du sang de la Côte d'Ivoire," *Transfusion* (Paris) 3, no. 1 (1960): 47–51; Rogoff to Director of Medical Services, October 29, 1962, BY4/69, KNA; "Malawi Red Cross Annual Report for 1964," box 6, Acc 0076/30(1), Malawi [Nyasaland until 1964], 1957–66, BRC London; "Territoire du Niger, Service de santé, 1959," box 64, IMTSSA; Federal Republic of Nigeria, *Annual Report of the Federal Ministry of Health, Nigeria,* January 1—December 31, 1960; Northern Nigeria, "Annual Report of Ministry of Health," 1965; République du Sénégal, Ministère de la santé publique, *Activité du Service,* 1969; Sierra Leone, *Report on the Medical and Health Services,* 1959; Republic of Tanganyika [Tanzania], Ministry of Health, "Annual Report," 1964; Uganda, Ministry of Health, *Annual Report,* 1963; Reports of British Red Cross Overseas Branches, 1947–1966, Acc 0106, BRC London.

Chapter 3: Blood Transfusion in Independent African Countries

1. Etienne Ayité, "La transfusion sanguine en Afrique noire de langue française" (Medical thesis, Dakar Faculté de médecine et de pharmacie, 1974), 32–33.

2. Unsigned copy of speech, almost certainly by Linhard, at dedication on December 2, 1964, unclassified records, Centre national de transfusion sanguine, Dakar, hereafter CNTS Dakar.

3. Uganda, Ministry of Health, Annual Report, July 1, 1961–June 30, 1962, 33–34; République du Sénégal, Ministère de la santé, *Rapport annuel, 1961,* 6–7; Kenya, Colony and Protectorate, *Ministry of Health Annual Report,* 1962, 44–45.

4. Richard M. Titmuss, *The Health Services of Tanganyika: A Report to the Government* (London: Pitman Medical Publishers, 1964), 34; Ralph Schram, *A History of the Nigerian Health Services* (Ibadan: Ibadan University Press, 1971), as cited in Adetokunbo A. Lucas, "What We Inherited: An Evaluation of What Was Left Behind at Independence and Its Effects on Health and Medicine Subsequently," in *Health in Tropical Africa during the Colonial Period,* ed. E. E. Sabben-Clare, David J. Bradley, and Kenneth Kirkwood (Oxford: Clarendon Press, 1980), 239–48.

5. On Ivory Coast, see Baba Sy, "Fonctionnement de la banque du sang de la Côte d'Ivoire," *Transfusion* 3, no. 1 (1960): 47–51; Ayité, "Transfusion sanguine," 34; D. Kerouedan et al., "Réflexions sur la transfusion sanguine en Afrique au temps de l'épidémie de sida: État des lieux et perspectives en Côte d'Ivoire," *Santé* 4, no. 1 (1994): 37–42. For Zambia, see Miss Jane Knudtzon, "Final Report from Zambia: October 1966"; "Zambia Red Cross Society," brochure (June 1966), Acc0076/61(1), Zambia (correspondence and reports, 1957–66), British Red Cross Museum and Archives, London, hereafter BRC London; Graco Matoka to Z. S. Hantchef, March 23, 1966, on the blood donor service in Zambia, AO985, International Federation of Red Cross and Red Crescent Societies, Geneva, Switzerland, hereafter IFRC; Republic of Zambia, *Ministry of Health Annual Report, for 1965 and 1966,* 46, 60; *Ministry of Health Annual Report for 1972,* 28–29.

6. Ayité, "Transfusion sanguine"; République du Tchad, Ministère de la santé publique, *Annuaire de statistique sanitaire, 1976,* 82; Kenya Medical Department, *Annual Report,* 1966, 154; Sierra Leone, *Report on the Medical and Health Service,* 1959 and 1970; Institut Pasteur du Cameroun, *Rapport annuel,* 1960–62, 69.

7. On the start of this process, which began with Truman's Point Four Program, based on his 1949 inaugural address, see Gilbert Rist, *The History of Development: From Western Origins to Global Faith,* 3rd ed., translated by Patrick Camiller (1996; trans., London: Zed, 2009), 69–79.

8. Republic of Kenya, *African Socialism and Its Application to Planning in Kenya,* Sessional Paper no. 10 (Nairobi, 1965), commented on in Bethwell A. Ogot and William Robert Ochieng', *Decolonization and Independence in Kenya, 1940–93* (Athens: Ohio University Press, 1995), 132–36. See also, Germano Mwabu, "Health Care Reform in Kenya: A Review of the Process," *Health Policy* 32, nos. 1–3 (April–June 1995): 245–55; George O. Ndege, *Health, State, and Society in Kenya* (Rochester: University of Rochester Press, 2001), 140–41.

9. Ndege, *Health, State,* 80–81.

10. John Iliffe, *East African Doctors: A History of the Modern Profession* (Cambridge: Cambridge University Press, 1998), 174–75.

11. For two snapshots of transfusion services in East Africa, see the following reports by Red Cross and WHO officials: "Visit by Dr. K. L. G. Goldsmith to Kenya,

acting as a Temporary Consultant for the World Health Organization, 16th–22nd October and 1st November 1972," AO658/2, IFRC; Z. S. Hantchef, "Mission to the Red Cross Societies of Ethiopia, Tanzania and Kenya," January 6–27, 1974, AO917 Tanzania, IFRC.

12. John A. Carman, *A Medical History of the Colony and Protectorate of Kenya: A Personal Memoir* (London: Rex Collings, 1976), 83–84; David Collins et al., "Hospital Autonomy: The Experience of Kenyatta National Hospital," *International Journal of Health Planning and Management* 14, no. 2 (1999): 129–53.

13. "Visit by Dr. K. L. G. Goldsmith to Kenya, 1974," IFRC.

14. "Blood Donation—The Jomo Kenyatta Blood Donation Week," editorial, *Medicus* 3, no. 11 (November 1984): 1.

15. Records and correspondence about Kenyatta Day exist at the Kenya National Blood Transfusion Service (hereafter cited as KNBTS) going back to 1984, and a newspaper clipping file from earlier. See, for example, reports from Kenyatta Day 1969: "Minister Donates Blood," *East African Standard,* October 15, 1969; "Hundreds Give One Pint Each," *Daily Nation,* October 15, 1969; "Blood Donors Set Record," October 25, 1969 [newspaper not identified].

16. Colony and Protectorate of Kenya, *Ministry of Health Annual Report,* 1959; Rogoff to Director of Medical Services, October 29, 1962, BY4/69, KNA; Republic of Kenya, *Ministry of Health Annual Report,* 1965. See also Republic of Kenya, "Country-wide Annual Report—Blood Donation—1981," June 11, 1982; Kenya Ministry of Health, "Memorandum on 1984 Kenyatta Day Celebration (32nd)," October 3, 1984; Republic of Kenya, Blood Donation Annual Reports from Provinces and Nairobi, 1986–88; Blood Donor Service, Nairobi, "Country-wide Blood Donation Report for the Year 1988," February 8, 1990, all in KNBTS.

17. Records of National Blood Transfusion Service, Nairobi, Kenya, 1984–2005.

18. N. Clumeck et al., "Acquired Immune Deficiency Syndrome in Black Africans," *Lancet* 1, no. 8325 (March 19, 1983). This article reported on African patients in Belgium. The first diagnosis of HIV/AIDS in Africa was the following year. See, for example, P. van de Perre et al., "Acquired Immunodeficiency Syndrome in Rwanda," *Lancet* 2, no. 8394 (July 14, 1984): 62–65.

19. More on this difficulty is found in the next section, but on the so-called "structural adjustments" in Kenya that responded to World Bank calls for cutting government spending, see Alfred Anangwe, "Health Sector Reforms in Kenya: User Fees," in *Governing Health Systems in Africa,* ed. Martyn Sama and Vinh-Kim Nguyen (Dakar: Codesria, 2008), 44–59; Joseph Kipkemboi Rono, "The Impact of the Structural Adjustment Programmes on Kenyan Society," *Journal of Social Development in Africa* 17, no. 1 (2002).

20. Rift Valley Provincial General Hospital, Nakuru, "Units Blood Collected and Screened [1988]," March 13, 1989, KNBTS.

21. Iliffe, *East African Doctors,* 174–75.

22. Republic of Kenya, "Country-wide annual report—Blood donation—1981," June 11, 1982; Kenya Ministry of Health, "Memorandum on 1984 Kenyatta Day Celebration (32nd)," October 3, 1984; Republic of Kenya, Blood Donation Annual Reports from provinces and Nairobi, 1986–88; Blood Donor Service Nairobi, "Country-wide Blood Donation Report for the Year 1988," February 8, 1990, KNBTS.

23. Population figures from Population Statistics, accessed August 25, 2011, www.populstat.info.

24. For a recent description of "call-responsive" donors at Kenyatta National Hospital (hereafter KNH), see F. Abdalla, O. W. Mwanda, and F. Rana, "Comparing Walk-in and Call-Responsive Donors in a National and a Private Hospital in Nairobi," *East African Medical Journal* 82, no. 10 (October 2005): 531–35. The authors found that over 97 percent of the nonremunerated donors (8,563) at the two largest hospitals in Nairobi from April 1999 to March 2000 were "call-responsive" donors. This usually meant a donor arranged by the family of someone needing a transfusion.

25. Iliffe provides examples of reports in the press and elsewhere of deteriorating conditions at KNH, including broken water taps, nonfunctioning toilets, out-of-stock supplies, lack of equipment maintenance, and drug racketeering. Iliffe, *East African Doctors,* 174–75.

26. Kenya, Minister of Health, memo, June 10, 1985, KNBTS.

27. Testing began at the end of 1988, according to annual reports in the central region, and most or all blood was screened in all other regions by the following year.

28. The exception is, perhaps, South Africa but that country is excluded from this study.

29. "Speech by His Excellency the Governor at the annual general meeting of the Uganda Red Cross Society, Kampala, 24th April 1952," Acc 0076/58(1) Uganda, 1949–56, BRC London.

30. D. J. Alnwick, M. R. Stirling, and G. Kyeyune, "Population Access to Hospitals, Health Centres and Dispensary/Maternity Units in Uganda, 1980," in *Crisis in Uganda: The Breakdown of Health Services,* ed. Cole P. Dodge and Paul D. Wiebe (Oxford: Pergamon Press, 1985), 194.

31. Florence Thomas, "Report on Uganda Red Cross Society," March 1966, Acc0287/63 Uganda, BRC London.

32. See chapter 5, "Who Gave Blood?"

33. *Uganda Red Cross Annual Reports,* 1956–1978 and Uganda Central Council Red Cross Branch, Annual reports, 1956–63, Acc 0076/58(1) Uganda, 1949–56, Acc 0076/58(2) Uganda, BRC London; Uganda Blood Transfusion Service records, Kampala (hereafter cited as UBTS), "Rapid Strengthening of Blood Transfusion Service in Uganda," proposal to PEPFAR, 2004; UBTS ; "Rapid Strengthening of Blood Transfusion Service in Uganda, End of Year 1 report [to PEPFAR]," April 2006, UBTS.

Sources for the period through the 1970s are from the Uganda Red Cross annual reports, supplemented by unpublished reports in the British Red Cross archives. Figures after 1989 are only for blood provided by the new Nakasero Blood Bank, as it was phased in initially to serve a region of the country up to one hundred kilometers from the capital, and then expanded to serve the whole country. It seems logical to conclude that hospital-based collection gradually declined as the national service grew.

For background, see E. J. Watson-Williams and P. K. Kataaha, "Revival of the Ugandan Blood Transfusion System 1989: An Example of International Cooperation,"

Transfusion Science 11, no. 2 (1990): 179–84; Rex Winsbury, ed., *Safe Blood in Developing Countries: The Lessons from Uganda* (Brussels: European Commission, 1995).

34. Alnwick, Stirling, and Kyeyune, "Population Access," 184; Stanley Scheyer and David Dunlop, "Health Services and Development in Uganda," in Dodge and Wiebe, *Crisis in Uganda,* 34; Iliffe, *East African Doctors,* 136–40, 150–52.

35. Juhani Leikola, "Report of a Mission to Uganda," November 10–14, 1984, Leikola, personal papers, Finnish Red Cross Blood Service, Helsinki, hereafter FRCBCS.

36. Private papers of Dr. Esau Nzaro, former head of hematology and blood bank at Mulago Hospital.

37. Leikola, "Report of a Mission."

38. Ibid.

39. Uganda Red Cross Society, *Annual Report for 1982/3,* 7.

40. See division reports for Jinja, Mpigi, Kampala North, and Kabale, in Uganda Red Cross Society, *Annual Report for 1984/5/6.*

41. Watson-Williams and Kataaha, "Ugandan Blood Transfusion"; Winsbury, *Safe Blood,* 59–62.

42. Winsbury, *Safe Blood,* 59.

43. Lieve Fransen, "Mission Report on trip to Uganda–Kenya 6–15 May 87," UBTS.

44. République du Sénégal, Ministère de la santé publique, *Activité du Service,* année 1961, 6; ibid., 1969, 10.

45. André Prost, *Services de santé en pays africain: Leur place dans des structures socio-économiques en voie de développement* (Paris: Masson, 1970), 55, 76–77; Maghan Keita, *A Political Economy of Health Care in Senegal* (Leiden, Brill: 2007), 46, 152–54. See also Léon Lapeyssonnie, *La médicine coloniale: Mythes et réalités* (Paris: Seghers, 1988). Dropout rates of between 20 and 25 percent were reported in the early years of the Dakar Faculty of Medicine. In 1976 24 out of 56 faculty members at the teaching hospital of the medical school were from France.

46. "[Jacques] Linhard Speech at inauguration of new CNTS building," December 2, 1964, Dakar, CNTS Dakar.

47. CNTS du Sénégal, "Activités pendant l'année 1961"; République du Sénégal, Ministère de la santé publique, "Statistiques sanitaires," 1961–69, CNTS Dakar.

48. Dried plasma declined from over eight thousand 350 cc bottles in 1965 (versus over ten thousand of whole blood) to less than a thousand in the early 1970s (there were no reports by the late 1970s). République du Sénégal, Ministère de la santé publique, "Statistiques sanitaires," 1965; 1970; 1972; 1974, CNTS Dakar. Guy Charmot, who served in the Dakar health service in the 1950s, reported problems with the purity of dried plasma even then. Charmot, interview by author, phone interview, Paris June 6, 2006.

49. B. A. Akué, [title unknown] (MD thesis, Université de Dakar, Faculté de médecine et de pharmacie, 1981), 27–28; Sénégal, Ministère de la santé publique, "Statistiques sanitaires," 1961–69, CNTS Dakar.

50. Figures from République du Sénégal, Ministère de la santé publique, *Statistiques sanitaires,* 1963–69; Akué, thesis, 39.

51. See figures 3.5 and 3.6 in next section.

52. These figures come from a more complete table.

TABLE 3.16. **Blood donations, Senegal, selected years, 1994–2003**

Hospital Blood Bank	*2003*	*2002*	*1994*	*Hosp. beds (1986)*
Tambacounda	1,345	1,220	432	115
Saint-Louis	—	1,230	1,297	266
Louga	882	573	455	110
Ndioum	516	187	75	114
Ouro Sogui	478	400	85	150
Thiès (Hôpital St. Jean de Dieu)	620	558	1,183	98
Thiès (region)	2,995	2,059	1,034	261
Kaolack	1,856	1,653	1,484	335
Diourbel	1,194	1,027	986	108
Ziguinchor	897	899	1,075	86
Kolda	694	—	294	0
Region subtotal	11,477	9,806	8,400	1,643
Dakar				
CNTS (incl. Hôpital le Dantec)	12,240	10,241	8,261	1,865
Hôpital principal	5,128	—	6,455	663
Total	**28,845**	**20,047**	**23,116**	**4, 171**

Sources: Drs. Kabou, Boyeldieu, "Rapport de Mission de Supervision des Postes de Transfusion Sanguine," February–March 1995; Bernard Poste, "Rapport de Mission de Supervision des Banques de Sang du Sénégal," March 2003; Bernard Poste, "Rapport de Mission de Supervision des Banques de Sang du Sénégal," June 2004, CNTS Dakar.

Some sources indicate donors and vary as to amount of donations. According to the reports, blood was sometimes collected in 250 ml and 450 ml bags, and occasionally in pint bottles (one UK pint is equal to 0.57 liters). On average, blood donations were slightly under 500 ml per donor.

53. Populstat, "Population Statistics: Historical Demography of All Countries, Their Divisions and Towns," accessed March 31, 2009, http://www.populstat.info.

54. For examples of the use of replacement donors, see Jean-Pierre Allain, Shirley Owusu-Ofori, and Imelda Bates, "Blood Transfusion in Sub-Saharan Africa," *Transfusion Alternatives in Transfusion Medicine* 6, no. 1 (March 2004): 16–23; D. Mignonsin et al., "Transfusion sanguine en Côte d'Ivoire: Perspectives d'avenir," *Médecine d'Afrique noire* 38, no. 11 (1991): 723–31; D. Nwagbo et al., "Establishing a Blood Bank at a Small Hospital, Anambra State, Nigeria," *International Journal of Gynecology and Obstetrics* 59, supp. 2 (1997): 135–39; Y. K. Sodahlon et al., "Sécurité transfusionnelle dans un contexte de ressources limitées: Processus de mise en place de la politique nationale transfusionnelle au Togo," *Cahiers d'études et de recherches francophones/Santé* 14, no. 2 (April–June 2004): 115–20.

55. For a partial overview of Nigeria, see Aba S. David-West, "Blood Transfusion and Blood Bank Management in a Tropical Country," *Clinics in Haematology* 10, no. 3 (October 1981): 1013–28. Information for the former Zaire is even more difficult to obtain, but according to Myriam Malengreau, who worked as a pediatrician between 1974 and 1989 in the lakes region around Katana, only hospitals in Kinshasa

and Lubumbashi had blood banks and a transfusion service. Elsewhere, hospitals took blood from family members as the need arose, guarding a dozen bottles or bags for emergencies. email Malengreau, pers. comm., February 19, 2006.

56. See, for example, Joan M. Nelson, ed., *Economic Crisis and Policy Choice: The Politics of Adjustment in the Third World* (Princeton: Princeton University Press, 1990).

57. An example is A. Edward Elmendorf, *Structural Adjustment and Health in Africa in the 1980s* (Washington, DC: American Public Health Association, 1993).

58. Joseph Kipkemboi Rono, "The Impact of the Structural Adjustment Programmes on Kenyan Society," *Journal of Social Development in Africa* 17, no. 1 (2002): 81–98; Anangwe, "Health Sector Reforms," 44–59.

59. Seydou Coulibaly and Moussa Keita, *Les comptes nationaux de la santé au Mali, 1988–1991* (Bamako: Institut national de recherche en santé publique, 1993), 147, as cited in Joseph Brunet-Jailly, "La santé dans quelques pays d'Afrique de l'Ouest après quinze ans d'ajustement," in *Crise et population en Afrique: Crises économiques, politiques d'ajustement et dynamiques démographiques,* ed. Jean Coussy and Jacques Vallin (Paris: Centre français sur la population et développement, 1996), 249, accessed June 23, 2012, http://www.ceped.org/cdrom/integral_publication_1988_2002/etudes/pdf/etudes_cpd_13; République du Sénégal, Ministère de l'économie, des finances et du plan, direction de la planification et direction de la prévision et de la statistique, *Tableau de bord annuel de la situation sociale au Sénégal, 1991* (Dakar: Ministère de l'économie, 1991), as cited in Joseph Brunet-Jailly, "La santé dans quelques pays d'Afrique de l'Ouest après quinze ans d'ajustement," in *Crise et population en Afrique,* ed. Jean Coussy and Jacques Vallin (Paris: Centre français sur la population et développement, 1996), 249, accessed June 23, 2012, http://www.ceped.org/cdrom/integral_publication_1988_2002/etudes/pdf/etudes_cpd_13.pdf.

60. République du Sénégal, Ministère de la Santé Publique, *Statistiques sanitaires et démographiques du Sénégal, année 1977,* 8. See also Keita, *Political Economy,* 133.

61. Scheyer and Dunlop, "Health Services," 33. Uganda and Kenya both use the shilling as their basic monetary unit. Their values fluctuate independently.

62. Compiled from World Bank, *Accelerated Development in Sub-Saharan Africa: An Agenda for Action* (Washington, DC: World Bank, 1981), 185. For background, see, Nicolas van de Walle, *African Economies and the Politics of Permanent Crisis, 1979–1999* (Cambridge: Cambridge University Press, 2001); Robert H. Bates, *When Things Fell Apart: State Failure in Late-Century Africa* (Cambridge: Cambridge University Press, 2008).

63. On Hantchef, see Evelyn von Steffan, "Dr. Hantchef (1910–2002) in memoriam," *Donor Recruitment International* 86 (November 2002): 12. There are also biographical personnel dossiers for Hantchef in AO650, IFRC.

64. "First International Red Cross Seminar on Blood Transfusion, Rome, 2 September 1958," AO904–1 IFRC International Congresses, IFRC.

65. "Ad Hoc Working Group, Vienna, 1961," AO911, IFRC.

66. For an early statement of extending transfusion expertise to developing countries, see "Summary of Practices: Ad Hoc Group of Experts on Blood Transfusion, First Meeting, Vienna, April 25–26, 1961," 37–38, AO911, IFRC.

67. Hantchef to regional director of League in Abidjan, October 13, 1966, AO915, IFRC. See also G. Miller, Canadian Red Cross, to Hantchef, May 21, 1969, AO915,

IFRC; "Upper Volta: A Motor-Cycle Did It!" *Blood Donor Recruitment: Successful Methods,* Medico-social Documentation no. 32 (Geneva: LRCS, n.d.), 17; "Haute-Volta: Dons reçus dans le cadre du Programme de Développement 1963/73," AO735/2 Upper Volta, IFRC.

68. Hantchef to Charles K. Johnson, General Secretary of Croix-Rouge du Dahomey, December 30, 1970, AO915 Benin, IFRC.

69. Croix-Rouge Togolaise, "Rapport d'activité, 1973," AO917, IFRC.

70. "Proposal for 3-year development plan for Gambian Red Cross Society," 1971; Central Committee of the Swedish Red Cross to LRCS, September 16, 1974, AO915, IFRC.

71. Request from J. J. van der Werf, Medical Director, Association Médico-sociale Oecuménique de Boende (Zaire) to Hantchef, March 29, 1972; John R. Van Dyck, Director of Institut de médicine tropicale, Kinshasa, to Hantchef, July 10 1973, AO917 Zaire, IFRC.

72. See International Working Group of Red Cross Blood Transfusion Experts, "Progress Reports," for the following meetings: 2nd meeting, Tehran, October 29–30, 1973; 3rd meeting, Bern, September 26–27, 1974; 4th meeting, Helsinki, August 1, 1975; 5th meeting, Bonn, June 25–26, 1976, AO911, IFRC archives.

73. "Contributions to the Development Programs and Donations," Red Cross Blood Transfusion Expert Working Group, 5th meeting, Bonn, June 28–30, 1976, 5–7, AO911, IFRC. Dollar values in table 3.13 are based on the historical exchange rate of June 1976. Currate, "Historical Exchange Rates," accessed March 15, 2013, http://currate.com/historical-exchange-rates.php.

74. Alan F. Fleming, "Memoir on University College Hospital, Ibadan, and the University of Ibadan, December 1962–July 1966," October 3, 1983, Mss Afr s1872, no. 52, Oxford Rhodes Library, Oxford, hereafter ORL.

75. See Institut Pasteur du Cameroun, *Rapport annuel, Institut Pasteur du Cameroun,* 1960–69.

76. The following is based on three accounts by the organizer of the Ethiopian transfusion service: Cyril Levene to G. W. Miller, June 4, 1970; Levene, "Red Cross Society Blood Bank Report," July 1971, AO912, IFRC. For a fuller account in hindsight, see Cyril Levene, "Brief History of the Development of the Transfusion Service," *How to Recruit Voluntary Donors in the Third World?* (Geneva: LRCS, 1984), 22–28.

77. R. Ghose, "History of Blood Transfusion in Ethiopia," *Ethiopian Medical Journal* 31, no. 4 (April 1963): 208–12.

78. Levene to Stewart, n.d. (presumably September 1969), AO915 Ethiopia, IFRC.

79. Levene, *How to Recruit,* 23.

80. Levene, "Ethiopian Red Cross," 6.

81. Aysheshim Ashress to Asfaw Desta, March 3, 1973, AO912, IFRC.

82. Zarco S. Hantchef, "Mission to the Red Cross Societies of Ethiopia, Tanzania and Kenya, January 6–27, 1974; Hantchef, "Visit of Jacques Moreillon to Ethiopia," August 13, 1974; Hantchef, "Report on ERC's Blood Program" [no date]; Martin E. Perret, "Memorandum, 22 October 1974," all in AO915 Ethiopia, IFRC.

83. Levene, *How to Recruit,* 22–29.

84. "The Gift of Life," *Spotlight* (Geneva: League of Red Cross and Red Crescent Societies, 1990), 8. A 1998 report indicated almost thirty thousand units collected

annually, nearly 80 percent by the national transfusion service. See D. Massenet, G. Tesfaye, and B. Dandera, "La transfusion sanguine en Ethiopie," *Médecine tropicale* 58, no. 3 (1998): 307–8. For a population of 58 million, however, this is a low rate of 52 donations/100,000 inhabitants.

85. Elisabeth-Brigitte Schindler, *Le Centre de transfusion sanguine de la Croix Rouge de Burundi: Son organisation et ses activités* (Bern: Croix Rouge Suisse, 1976), 7. For earlier discussions, see "LCRS, Secteur opération, plan d'action 1968 du programme de développement, rapport trimestriel, juillet–septembre 1968," 2, AO917 Burundi, IFRC.

86. François Buyoya to Jean Pascalis, late May 1972, AO917 Burundi, IFRC.

87. "Premier rapport du Projet de Développement du Centre National de Transfusion Sanguine," Burundi 1974, AO917, IFRC.

88. Schindler, *Centre de transfusion,* 8.

89. Proposal from the president of the Rwandan Red Cross to the LRCS about Red Cross assistance, April 6, 1971; Madame Egger of the LRCS to the president of the Croix-Rouge Rwandaise, April 29, 1971, both in AO916, IFRC.

90. "Note à Monsieur Vercamer," August 28, 1972; Vercamer and R. Vermeylan, Administrateur General of Belgian Red Cross to G. Cuddell, Belgian Red Cross, January 29, 1974; Projet Inter-Croix-Rouge de transfusion sanguine au Rwanda, box 7/30, DIA 7/287, Archives of Belgian Red Cross, Mechelen, hereafter BRC Mechelen.

91. Ministère de la Santé publique et des affaires sociales, "Plan générale de development des service médicaux-sociaux, 1975/80, section B, Unité de transfusion sanguine à Kigali," Internationale Acties 7/22 CTS Kigali, BRC Mechelen.

92. Carl Vandekerckhove, Director General, Belgian Red Cross, to Monsieur De Vleeschouwer, AGCD, "Rapport d'évaluation de la transfusion sanguine au Rwanda," July 31, 1985, Vandekerckhove, 1978–85, Allerlei documentation, BRC Mechelen.

93. Ibid. See also "Projet Inter-Croix-Rouge de transfusion sanguine au Rwanda, Rapport intérimaire," 1984, 1986, BRC Mechelen.

94. "Rapport intérimaire de l'exercice 1992," Projet Inter-Croix-Rouge de transfusion sanguine au Rwanda, Rapport intérimaire, DIZ-DIA 5/14 Project Dossiers BTC Rwanda 91–92–93, BRC Mechelen.

95. Jukka Koistenan and Juhani Leikola, interview by author, Helsinki, July 6, 2006.

96. There are extensive archives at the Finnish Red Cross Blood Service in Helsinki. For example, on financing the initial Somalia project, see D. B. L. McClelland and E. Linnakko, "Evaluation of Somalia Red Cross Blood Transfusion Service Development Project, 1986–1988," November 1989, FRC. See also "International Red Cross Cooperation in Blood Transfusion," 1984, AO912, IFRC.

97. See, for example, an assessment visit sponsored by WHO to Gabon by Linhard, the long-serving director of the Dakar transfusion center, "Rapport de Mission de Linhard, February 15–28, 1974," WHO library archives, Geneva.

98. *Premier congrès africain de transfusion sanguine, 7–12 Mars 1977, Yamoussoukro, Côte d'Ivoire, actes du congrès* (Paris: Institut INNIT, 1981). There is little about the history of transfusion in the Ivory Coast, but some information can be found in Sy, "Banque du Sang"; Kerouedan et al., "Transfusion sanguine." A periodic bulletin, *Transfusion ivoirienne,* began publication in 1992.

99. Of the references already cited, see especially Rex Winsbury, ed., *Safe Blood in Developing Countries: The Lessons from Uganda* (Brussels: European Commission, 1995).

100. Blood Donor Services, Nairobi, "Annual Report for the Year 1998," January 15, 1999, Archives of Kenya National Blood Transfusion Service, Nairobi, hereafter KNBTS. Internal memos show the inability of the Nairobi service to handle the large numbers of volunteers for blood donation (e.g., students at Nairobi University) or store the volume of blood donated. James A, Mwalloh to head, National Public Health Laboratory Service, October 30, 1998, KNBTS.

101. Jack Nyamongo to Director of Medical Services, Kenya Ministry of Health, "Re: Committee to develop the Kenya Blood Transfusion Policy," memorandum, August 2, 1999, KNBTS. After the bombing, the United States had offered to send blood to Kenya. See also Nyamongo to Dr. Chotara, WHO representative, Kenya, "Re: Blood Donation towards Bomb Blast," October 30, 1998, KNBTS. For background, see Saade Abdallah, Rebekah Heinzen, and Gilbert Burnham, "Immediate and Long-Term Assistance Following the Bombing of the U.S. Embassies in Kenya and Tanzania," *Disasters,* 31, no. 4 (December 2007): 417–34.

102. Initial-year U.S. congressional funding announced for PEPFAR in 2004 was over $54 million to eleven countries in Africa, just for blood safety. United States, Department of State, Office of the U.S. Global AIDS Coordinator, "The President's Emergency Plan for AIDS Relief: Report on Blood Safety and HIV/AIDS," June 2006, accessed September 2, 2011, http://www.state.gov/documents/organization/74125.pdf.

Chapter 4: Who Got Blood?

1. Examples of studies on use and misuse are R. I. Anorlu et al., "Uses and Misuse of Blood Transfusion in Obstetrics in Lagos, Nigeria," *West African Journal of Medicine* 22, no. 2 (June 2003): 124–27; Benedik R. Holzer et al., "Childhood Anemia in Africa: To Transfuse or Not Transfuse?" *Acta tropica* 55, no. 1 (October 1993): 47–51; J. R. Zucker et al., "Anaemia, Blood Transfusion Practices, HIV and Mortality among Women of Reproductive Age in Western Kenya," *Transactions of the Royal Society for Tropical Medicine and Hygiene* 88, no. 2 (March–April 1994): 173–76; and chapter 6, this volume.

2. Émile Lejeune, "Transfusion sanguine après hémoglobinurie grave," *Annales de la Société belge de médecine tropicale* 1 (1921): 299–300; D. Spedener, "Le traitement des pneumonies des noirs par transfusion de sang des convalescents," *Bulletin médical du Katanga* 1 (1924): 234–38; Germond, "Statistiques des cas de pneumonie traités par transfusion de sang de convalescents," *Bulletin médical du Katanga* 1 (1924): 243. Another early transfusion reported in Southern Rhodesia was for blackwater fever. See S. G. Gasson, "The First Blood Transfusion in Southern Rhodesia," *Central African Journal of Medicine* 6 (November 1960): 490.

3. A. Lodewyck, "Note sur la transfusion sanguine chez les nourrissons et les enfants," *Recueil de travaux de sciences médicales au Congo belge* 2 (1944): 157–61; C.-S. Ronsée, "Anémies malariennes des enfants et transfusions sanguines, avec observations sur les groupes sanguins des Bakongo," *Mémoires Institut royal colonial Belge, Section des sciences naturelles et médicales* 20, no. 2 (1952): 1–64; M. P. de Smet, "Traitement d'attaque des anémies-oedèmes graves par transfusions fractionnées chez des enfants sevrés," *Annales de la Société belge de médecine tropicale* 34, no. 2 (1954): 155–69.

4. C. Bouyer, "Perfusions et transfusions par la veine sous-clavière chez les nourissons et les enfants," *Annales de la Société belge de médecine tropicale* 45, no. 1 (1965): 39–48.

5. Institut Pasteur de Brazzaville (Congo), "Rapport sur le fonctionnement technique de l'Institut Pasteur de Brazzaville," 1955–56, IPO-RAP, box 11, Archives de l'Institut Pasteur, Paris.

6. Musée international de la Croix-Rouge, Geneva, archives (hereafter cited as MICR), poster collection. I am grateful to the curator, Sophie Chapuis, for her generous assistance.

7. BBT-1988-107-13_A003GU, MICR.

8. Una Maclean, "Blood Donor Recruitment in Ibadan: The Record of One Year's Experience," *Journal of Tropical Medicine and Hygiene* 61, no. 12 (1958): 311–14; Maclean, "Blood Donors for Nigeria," *Community Development Bulletin* 11 (1960): 26–31.

9. Maclean, "Blood Donors for Nigeria," 29.

10. BBT-2003-35-20_a064jb, MICR. This same poster was also used in Liberia in 1970. BBT-2002-29-75_a023jb.

11. BBT-2002-24-88_a016jb; BBT-2002-16-15_a009jb, MICR.

12. BBT-2002-27-33_a017jb, MICR.

13. BBT-2002-29-68_a023jb; BBT-2002-29-66_a023jb; BBT-2002-29-69_a023jb, MICR.

14. AO915, International Federation of Red Cross and Red Crescent Societies, hereafter IFRC. The date of the poster is September 19, 1973.

15. The black-and-white poster, BBT-2002-15-29_a007jb, is from MICR and has no date. The color poster is dated around 1980 and is in 53 Angola Blutspendedienst finanzelle, korrespond., 1979–81, Swiss Red Cross archives, Bern, hereafter SRC. Presumably both posters were created as part of Swiss Red Cross assistance to Angola beginning in 1978. See chapter 3.

16. BBT-2002-16-24_a009jb, MICR.

17. BBT-2002-27-62_a018jb, MICR.

18. *Le lien du sang* (1976), 12. The subtitle of the periodical is "Revue trimestrielle des Donneurs de Sang Bénévolés de la République Populaire du Bénin." The poster contest was held in primary schools for blood donation, and over five hundred entries were received. AO73611 Benin, IFRC.

19. O. Walter and L. Langlo, "A Blood-Bank Service in a Rural Hospital in East Africa," *East African Medical Journal* 39 (December 1962): 702–7.

20. M. Sankalé, H. Ruscher, and Y. Touré, "Accidents et incidents de la transfusion sanguine et leur prévention dans un service de médecine à Dakar," *Bulletin de la Société médicale d'Afrique noire de langue française* 18, no. 3 (1973): 307–9.

21. Elisabeth-Brigitte Schindler, *Le Centre de transfusion sanguine de la Croix-Rouge de Burundi: Son organisation et ses activités* (Bern: Croix-Rouge Suisse, 1976), 7–8.

22. Records supplied from personal papers of Professor Esau Nzaro, Senior Consultant, Mulago Blood Bank, Kampala, Uganda, May 2006.

23. Aba S. David-West, "Blood Transfusion and Blood Bank Management in a Tropical Country," *Clinics in Haematology* 10, no. 3 (October 1981): 1013–28. Before 1972 and after 1983 David-West published under the name A. S. Sagoe. At the 1977

African Congress of Blood Transfusion in the Ivory Coast, David-West presented three talks: "Blood Transfusion in Africa (English-Speaking Countries)," "Free Giving Blood—Socio-Cultural Problems," and "Utilisation of Blood at the University Hospital-Ibadan."

24. "Relève mensuel des groupages sanguins, 1967–1979," 1980, box 53 Obervolta /Blutspendedienst 1976–84, SRC.

25. "Dons reçus dans le cadre du Programme de Développement 1963/73," AO735/2 Upper Volta, IFRC. The list included Red Cross societies of Canada, the United States, the Soviet Union, West Germany, France, and Luxembourg. For a brief history, see Joseph Conombo, "Croix-Rouge voltaïque," *Don universel du sang* 22 (1972): 19–20.

26. "Relève mensuel des groupages sanguins, Hôpital Yalgado Ouedragogo, 1967–79"; see also Evelyn Steffens, "Rapport de Mission à Ouagadougou, Burkina Faso, 23–28 October, 1984," both in box 53, Obervolta/Blutspendedienst 1976–84, SRC. There is no explanation for the large percentage of "Divers demands."

27. A. O. Emeribe et al., "Blood Donations and Patterns of Use in Southeastern Nigeria," *Transfusion* 33, no. 4 (1993): 330–32.

28. In fact, the article, published in 1993, makes no mention of HIV/AIDS.

29. Alan E. Greenberg et al., "The Association between Malaria, Blood Transfusions, and HIV Seropositivity in a Pediatric Population in Kinshasa, Zaire," *Journal of the American Medical Association* 259, no. 4 (1988): 545–49.

30. Most figures come from Projet Inter-Croix-Rouge de Transfusion Sanguine au Rwanda, "Rapport annuel," 1976–93, Belgian Red Cross, Mechelen, hereafter BRC Mechelen.

31. Pediatric use continues to predominate in many countries of sub-Saharan Africa today. See chapter 6 for the controversy it has produced.

32. Sandrine Simeu Kamdem, "Evolution de la pratique de la transfusion sanguine dans 3 hôpitaux du Cameroun" (medical thesis, University of Yaoundé, 2010).

33. Ibid.

34. Ibid.

35. E. O. Addo-Yobo and H. Lovel, "How Well Are Hospitals Preventing Iatrogenic HIV? A Study of the Appropriateness of Blood Transfusions in Three Hospitals in the Ashanti Region, Ghana." *Tropical Doctor* 21, no. 4 (1991): 162–64. As mentioned in chapter 3, note 52, units could vary between 250 and 500 ml. Even a 500 ml unit is considered ineffective for most adult patients as a rule of thumb.

36. This article ushered in a whole series of similar studies. In addition to Holzer et al., "Childhood Anemia" and Anorlu et al., "Uses and Misuse," see J. Vos et al., "Are Some Blood Transfusions Avoidable? A Hospital Record Analysis in Mwanza Region, Tanzania," *Tropical and Geographical Medicine* 45, no. 6 (1993): 301–3.

37. D. Mignonsin et al., "Transfusion sanguine en Côte d'Ivoire: Perspectives d'avenir," *Médecine d'Afrique noire* 38, no. 11 (1991): 723–31.

38. Eve M. Lackritz et al., "Blood Transfusion Practices and Blood-Banking Services in a Kenyan Hospital," *AIDS* 7, no. 7 (1993): 995–99; F. R. Barradas, T. Schwalbach, and A. Novoa, "Blood and Blood Products Usage in Maputo," *Central African Journal of Medicine* 40, no. 3 (1994): 56–60; Dora Mbanya, Fidele Binam, and Lazare

Kaptué, "Transfusion Outcome in a Resource-Limited Setting of Cameroon: A Five-Year Evaluation," *International Journal of Infectious Diseases* 5, no. 2 (2001): 70–73.

39. For examples from both ends of the historical spectrum, see G. MacDonald, "Theory of the Eradication of Malaria," *Bulletin of the World Health Organization* 15, nos. 3–5 (1956): 369–87; Ricardo Águas et al., "Prospects for Malaria Eradication in Sub-Saharan Africa," *PLoS One* 12, no. 3 (March 2008): e1767, accessed September 29, 2011, http://www.ncbi.nlm.nih.gov/pmc/articles/PMC2262141/.

Chapter 5: Who Gave Blood?

1. See, for example, Mwelwa C. Musambachime, "The Impact of Rumor: The Case of the Banyama (Vampire Men) Scare in Northern Rhodesia, 1930–1964," *International Journal of African Historical Studies* 21, no. 2 (1988): 201–15; and the more extensive study by Luise White, *Speaking with Vampires: Rumor and History in Colonial Africa* (Berkeley: University of California Press, 2000). But overall these works are exceptions.

2. White, *Speaking with Vampires,* 16, 106–8, 126–28. In the course of research for this book, occasional references were made to such rumors about blood by some staff of blood banks in African hospitals, but it was not possible to do a systematic examination of this quite important question. Personal communication, O. W. Mwanda, Kenyatta National Hospital, May 4, 2006.

3. Emile Lejeune, "Transfusion sanguine après hémoglobinurie grave," *Annales de la Société belge de médecine tropicale* 1 (1921): 299–300.

4. D. Spedener, "Le traitement des pneumonies des noirs par transfusion de sang des convalescents," *Bulletin médical du Katanga* 1 (1924): 234–38.

5. Joseph Lambillon and N. Denisoff, "Étude de l'organisation d'un service de transfusions sanguines dans un centre hospitalier d'Afrique," *Annales de la Société belge de médecine tropicale* 20 (1940): 279–85.

6. Ibid., 284.

7. Jean Langeron, "Étude sur la sérotherapie de la pneumonie: Valeurs thérapeutiques des sérums des convalescents," *Bulletin médical du Katanga* 1 (1924): 231.

8. George Valcke, "Note," *Annales de la Société belge de médecine tropicale* 14 (1934): 432–33.

9. K. S. Dewhurst, "Observations on East African Blood Donors," *East African Medical Journal* 22 (1945): 276–78.

10. Gaston Ouary, "Compte rendu de la Mission effectuée par le Médecin Commandant Ouary au Centre de Transfusion d'Alger (15 juillet–15 octobre 1944)," series H, 1H1 (1), Archives nationales du Sénégal, hereafter ANS.

11. Ibid., 20.

12. Ibid., 13–14.

13. Ibid., 15.

14. For background, see Edouard Benhamou, "Notes pour servir à l'histoire de la transfusion sanguine dans l'armée française de 1942 à 1945 à partir de l'Afrique du Nord," *Revue des Corps de santé des Armées Terre, Mer, Air* 7, "special" (November 1966): 859–62.

15. Institut Pasteur (Senegal),"Rapport sur le fonctionnement technique de l'Institut Pasteur de Dakar," 1945, 23–24.

16. Institut Pasteur (Senegal),"Rapport sur le fonctionnement technique de l'Institut Pasteur de Dakar," 1948, 22–23.

17. "Arrêté no. 465, 28 avril 1951, Gouvernement Général de l'Afrique Occidentale, Direction générale de a Santé publique," series H, 1H62 (163), textes officiels, ANS.

18. Jacques Linhard, "Le centre fédéral de transfusion de l'AOF," *Médecine tropicale* 11 (1951): 957.

19. Ibid.

20. Croix-Rouge Togolaise, "Rapport d'activité 1973," AO917, International Federation of Red Cross and Red Crescent Societies, hereafter IFRC; "13 Sept 1957 Assemblé Territoriale de la Côte d'Ivoire projet de déliberation tendant à la création en Côte d'Ivoire d'un Centre de Transfusion Sanguine," box 294, Institut de médecine tropicale du service de santé des armées, Marseille, hereafter IMTSSA. The Upper Volta Red Cross reported that in 1958 it was limited to a shipment of only four liters of blood a week flown to the capital at Ouagadougou from Dakar. See below, *Don universel du sang* 22 (1972): 19–21.

21. William H. Schneider, "Blood Transfusion between the Wars," *Journal of the History of Medical and Allied Sciences* 58, no. 2 (April 2003): 187–224. For background on the French system, see Marie-Angèle Hermitte, *Le sang et le droit: Essai dur la transfusion sanguine* (Paris: Éditions de Seuil, 1996), 70–77.

22. "Miss Borley's Report," November 1950, box 1, Acc 0076/6(1) Basutoland, British Red Cross Museum and Archives, London, hereafter BRC London.

23. "Lady Limerick's Report," January 25–29, 1954, box 4, Acc 0076/21(1) Gambia, BRC London.

24. Amy Chipp, "Organiser's report for the months of March and April 1954," Acc 0076/36(5) Nigeria, Lagos and Colony, BRC London.

25. Christine Burton, "Eastern region report- May 1st, 1954 to June 4th, 1954," June 16, 1954, Acc 0076/36(4) Nigeria, Eastern Region, BRC London.

26. "Northern Rhodesia Branch, Director's and Treasurer's report to annual meeting," October 22, 1952, Acc 0076/38(1) Northern Rhodesia, BRC London.

27. *Nigerian Branch News Sheet,* no. 5 (October 1953), Acc 0076/36(1) Nigeria, BRC London.

28. "Western region of Nigeria, Monthly report for May and June 1953," Acc 0076/36(2) Nigeria, Western Region; *Nigerian Branch News Sheet,* no. 8 (July 1954), Acc 0076/36(1) Nigeria, BRC London.

29. I am grateful to Sophie Chapuis at the International Red Cross and Red Crescent Museum, Geneva, for providing a copy of the film.

30. "Organiser's report for the months of March and April 1954, Amy Chipp," Acc 0076/36(5) Nigeria, Lagos and Colony; "Western Region of Nigeria, Monthly Report for May 1953," Acc 0076/36(2) Nigeria, Western Region, BRC London.

31. "BRCS Kenya (Central Council) Branch Report for the year ending 31 October 1953," 22, Acc 0287/37 Kenya, BRCS. On Uganda, see "Uganda Branch of the BRCS," BRCS annual report, 1951–52, 2–3, Acc 0287/63 Uganda; Uganda Red Cross,

branch report, in *Report of the British Red Cross Society,* 1956, 63; ibid., 1958 report, 24, BRC London.

In 1963, Dr. Ian McAdam, one of the founders of the Ugandan transfusion service, proposed a film on the history of the service. See "Miss Whittington's Report on Visit to East Africa & Aden, Feb 28—March 29, 1963," Acc 0076/58(2) Uganda, BRC London. On Nyasaland, see "Director's report Oct-Dec 1961," box 6, Acc 0076/30(1) Malawi [Nyasaland until 1964], BRC London. On Zanzibar, see "Zanzibar Committee Annual Report December1962," Acc 0076/62(2) Zanzibar, BRC London.

32. Teresa Spens, "Notes on a Visit to East African Branches, Sept. to November 1953," Acc 0076/2(1), BRC London.

33. "West Africa: Lady Limerick's tour, January 23 to February. 19, 1961," 24, Acc0076/59(2), BRC London.

34. Patricia Jephson, "Nyasaland Report, February, March 1961," box 6, Acc 0076/30(1) Malawi [Nyasaland until 1964], BRC London. The British Red Cross was quite adamant about not being associated with payment for blood.

35. Jephson to Miss Whittington, June 3, 1961, box 6, Acc 0076/30(1) Malawi, BRC London.

36. "Nigeria 1956–65," 44 pp. of typescript manuscript, May 10, 1983, Mss Afr s 1872, no. 99, Oxford Rhodes Library, hereafter ORL.

37. Alan F. Fleming, "Memoir on University College Hospital, Ibadan, and the University of Ibadan, December 1962–July 1966," October 3, 1983, Mss Afr s1872, no. 52, ORL.

38. Una Maclean, "Blood Donor Recruitment in Ibadan: The Record of One Year's Experience," *Journal of Tropical Medicine and Hygiene* 61, no. 12 (1958): 311–12.

39. Una Maclean, "Nigeria 1956–65," p. 10, May 10, 1983, Mss Afr s1872, no. 99, ORL.

40. Maclean, "Blood Donor Recruitment," 313.

41. M. E. Enosolease, C. O. Imarengiaye, and O. A. Awodu, "Donor Blood Procurement and Utilisation at the University of Benin Teaching Hospital, Benin City," *African Journal of Reproductive Health* 8, no. 2 (August 2004): 59–63. Aba S. David-West reported in 1981, "Of the 19 states of Nigeria, only Oyo State has an organized blood transfusion service that distributes blood on demand to hospitals within the state." David-West, "Blood Transfusion and Blood Bank Management in a Tropical Country," *Clinics in Haematology* 10, no. 3 (1981): 1014.

42. "Blood Transfusion Service, sub-committee of BMA on Blood Transfusion Service, minutes of meeting held 4 May 1948 at Mulago Hospital, Kampala," Acc 0076/58(1) Uganda, BRC London.

43. *Don universel du sang* 22 (1972): 19–21.

44. "Desperate Need for Blood," *Sunday Nation* (Nairobi), December 12, 1971.

45. Uganda Red Cross Society, *Annual Report,* 1970, 31, Periodicals: Uganda, box 16723, IFRC.

46. Elisabeth-Brigitte Schindler has pictures of the inauguration, including the blood donations. Schindler, *Le Centre de transfusion sanguine de la Croix Rouge de Burundi: Son organisation et ses activités* (Bern: Croix Rouge Suisse, 1976). For examples in the Kenyan press of government officials donating blood, see "Minister Donates Blood,"

East African Standard, October 15, 1969; "Minister of Commerce," "Mayor of Nairobi," *East African Standard,* October 20 [1969?].

47. "Girls Lead Blood Drive in Mombasa," unlabeled clipping, 1969(?), Kenya National Blood Transfusion Service, Nairobi, hereafter KNBTS; "Blood . . . and Why You Need It," *Daily Nation* (Nairobi), October 22, 1973.

48. Private papers of Dr. Esau Nzaro, former head of hematology and the blood bank at Mulago Hospital, Kampala, Uganda; Juhani Leikola, "Report of a Mission to Uganda," November 10–14, 1984, Juhani Leikola personal papers, Finnish Red Cross Blood Service, Helsinki. See also chapter 3, this volume.

49. Professor Myriam Malengreau, École de santé publique Université catholique de Louvain, Brussels, pers. comm., February 9, 2006, "Quelques informations sur les transfusions à la Fomulac-Katana entre fin 1974 et 1995." Current practice in the United States is for donors to wait eight weeks between donations of whole blood, to avoid risk of anemia.

50. Uganda Red Cross, *Annual Report,* 1970, 2, Periodicals: Uganda, box 16723, IFRC.

51. Uganda Red Cross, *Annual Report,* 1975, 5, Periodicals: Uganda, box 16723, IFRC.

52. Didier Fassin, "Le domaine privé de la santé publique: Pouvoir, politique et sida au Congo," *Annales: Histoire, sciences sociales* 49, no. 4 (1994): 763–64.

53. "Nairobi Blood Donor Services, Annual Report—Blood Donation 1989," March 28, 1990, KNBTS.

54. "Nairobi Blood Donor Services, Annual Report—Blood Donation 1990," March 11, 1991, KNBTS.

55. Projet Inter-Croix-Rouge de Transfusion Sanguine au Rwanda, "Rapport annuel du travail," 1986 and 1987, Internationale Acties 7/24, Belgian Red Cross, Mechelen, hereafter BRC Mechelen.

56. See chapter 3; E. J. Watson-Williams and P. K. Kataaha, "Revival of the Ugandan Blood Transfusion System 1989: An Example of International Cooperation," *Transfusion Science* 11, no. 2 (1990): 179–84.

57. Dr. Nseka Kifuani, "La situation actuelle de la transfusion dans notre pays," *Hôpital africain* 15 (1983): 18.

58. Between 2004 and 2010 the Uganda Blood Transfusion Service received almost $24 million from PEPFAR, which was 65 percent of the total UBTS budget. Jerry P. Lunier, "Dedication of Uganda Blood Transfusion Service Facility," February 25, 2012, last accessed March 21, 2013, http://kampala.usembassy.gov/remarks_02252012.html.

59. Dr. Claudien Kamilindi to Vermeylan, October 25, 1979, Internationale Acties 7/23, BRC Mechelen.

60. Projet Inter-Croix-Rouge de Transfusion Sanguine au Rwanda, "Rapport annuel du travail," 1983, 1987, 1992, 1993, Internationale Acties 7/24, 7/31, 5/14, 5/22, BRC Mechelen.

61. "Rapport d'évaluation de la transfusion sanguine au Rwanda, Projet cofinancé par l'AGCD et la Croix-rouge de Belgique," July 31, 1985, 2, Internationale Acties 7/22, BRC Mechelen.

62. Projet Inter-Croix-Rouge de Transfusion Sanguine au Rwanda, "Rapport annuel du travail, 1985," Internationale Acties 7/24, BRC Mechelen.

63. Projet Inter-Croix-Rouge de Transfusion Sanguine au Rwanda, "Rapport annuel du travail, 1976–1993," Internationale Acties 7/23, 7/24, 7/31, 5/14, 5/22, BRC Mechelen.

64. Neelam Dhingra, "Making Safe Blood Available in Africa," statement before Committee on International Relations, Subcommittee on Africa, Global Human Rights and International Operations, U.S. House of Representatives, June 27, 2006, accessed April 30, 2010, http://www.who.int/bloodsafety/makingsafebloodavailableinafricastatement.pdf.

65. Jean-Pierre Allain, Shirley Owusu-Ofori, and Imelda Bates, "Blood Transfusion in Sub-Saharan Africa," *Transfusion Alternatives in Transfusion Medicine* 6, no. 1 (2004): 16–23.

66. I. Bates, G. Manyasi, and A. Medina Lara, "Reducing Replacement Donors in Sub-Saharan Africa: Challenges and Affordability," *Transfusion Medicine* 17, no. 6 (December: 2007): 435.

67. F. Abdalla, O. W. Mwanda, and F. Rana, "Comparing Walk-in and Call-Responsive Donors in a National and a Private Hospital in Nairobi," *East African Medical Journal* 82, no. 10 (October 2005): 531–35.

Chapter 6: Blood Transfusion and Health Risk before and after the AIDS Epidemic

1. P. van de Perre et al., "Antibody to HTLV-III in Blood Donors in Central Africa," *Lancet* 1, no. 8424 (February 9, 1985): 336–37; Jonathan M. Mann et al., "Surveillance for AIDS in a Central African City. Kinshasa, Zaire," *Journal of the American Medical Association* 255, no. 23 (June 20, 1986): 3255–59. For an excellent summary, see T. Quinn et al., "AIDS in Africa: An Epidemiological Paradigm," *Science* 234, no. 4779 (November 21, 1986): 955–63.

2. See below, and for an early study with some historical perspective, see Alan F. Fleming, "HIV and Blood Transfusion in Sub-Saharan Africa," *Transfusion Science* 18, no. 2 (June 1997): 167–79. For more recent assessments, see the report of the PEPFAR program that includes twelve African countries. Jerry A. Holmberg et al., "Progress toward Strengthening Blood Transfusion Services—14 Countries, 2003–2007," *Morbidity and Mortality Weekly Report* 57, no. 47 (November 28, 2008): 1273–77; S. Jayaraman et al., "The Risk of Transfusion-Transmitted Infections in Sub-Saharan Africa," *Transfusion* 50, no. 2 (February 2010): 433–42; M. van Hulst, C. T. Smit Sibinga, and M. J. Postma, "Health Economics of Blood Transfusion Safety—Focus on Sub-Saharan Africa," *Biologicals* 38, no. 1 (January 2010): 53–58.

3. There has been very little study of this question, but for one broad overview, see Harvey J. Alter and Harvey G. Klein, "The Hazards of Blood Transfusion in Historical Perspective," *Blood* 112, no. 7 (October 2008): 2617–26, which focuses on the discovery of hepatitis B. For general background, see Susan E. Lederer, *Flesh and Blood: Organ Transplantation and Blood Transfusion in Twentieth-Century America* (New York: Oxford University Press, 2008).

4. For background, see A. D. Farr, "The First Human Blood Transfusion," *Medical History* 24, no. 2 (1980): 143–62; Kim Pelis, "Blood Clots: The Nineteenth-century Debate over the Substance and Means of Transfusion in Britain," *Annals of Science* 54, no. 4 (1997): 331–60.

5. On the beginning of the modern era of blood transfusion after 1900, see William H. Schneider, "Blood Transfusion in Peace and War, 1900–1918," *Social History of Medicine* 10, no. 1 (April 1997): 105–26.

6. F. Oehlecker, "Erfahrungen aus 170 direkten Bluttransfusionen von Vene zu Vene," *Archiv für klinische Chirurgie* 116 (1921): 714–15.

7. William H. Schneider, "Blood Transfusion between the Wars," *Journal of the History of Medicine and Allied Sciences* 58, no. 2 (April 2003): 187–224.

8. In addition to Alter and Klein, "Hazards of Blood Transfusion," for France, see Ann-Marie Casteret, *L'affaire du sang* (Paris: La Découverte, 2002); for America, see Judith Reitman, *Bad Blood: Crisis in the American Red Cross* (New York: Kensington Books, 1996).

9. Indiana Blood Center (Indianapolis), communication with author, January 12, 2005. For current guidelines, see United States, Department of Health and Human Services, U.S. Food and Drug Administration, "Keeping Blood Transfusions Safe: FDA's Multi-layered Protections for Donated Blood," accessed July 8, 2013, http://www.fda.gov/BiologicsBloodVaccines/SafetyAvailability/BloodSafety/ucm095522.htm.

10. Alter and Klein, "Hazards of Blood Transfusion"; Baruch S. Blumberg, *Hepatitis B: The Hunt for a Killer Virus* (Princeton: Princeton University Press, 2003). In some ways foreshadowing HIV, the HBV was newly detected and more widely spread by modern medical techniques such as injection and transfusion, but it is a much older virus, perhaps 30,000 years old. See Dimitrios Paraskeuis et al., "Dating the Origin and Dispersal of Hepatitis B Virus Infection in Humans and Primates," *Hepatology* 57, no. 3 (March 2013): 908–16.

11. Juhani Leikola, "How Much Blood for the World?" *Vox sanguinis* 54, no. 1 (1988): 2.

12. Émile Lejeune, "Transfusion sanguine après hémoglobinurie grave," *Annales de la Société belge de médecine tropicale* 1 (1921): 299–300.

13. For example, in 1954 the Central Pathology Laboratory at the main hospital (Sewa Hadji Hospital) in Dar es Salaam, Tanganyika, reported almost thirty-six hundred Kahn tests for syphilis during the year. Tanganyika Territory, *Annual Report of the Medical Laboratory, 1954,* 10.

14. G. M. Edington, "Some Observations on Blood Transfusion in the Gold Coast," *West African Medical Journal* 5, no. 2 (1956): 72, 74–75.

15. Clapham report on setting up the Ghana Red Cross Society, April 29, 1958, Acc 0076/24(1) Gold Coast, British Red Cross Museum and Archives, London, hereafter BRC London.

16. For the thirtieth anniversary of the beginning of the blood transfusion center in Senegal, see the special issue of *Bulletin de la Société médicale d'Afrique noire de langue française* 18 (1973).

17. "Rapport de Mission de Linhard, 15–28 Fev 1974," WHO AFR/HLS/27, World Health Organization archives, Geneva.

18. J. Linhard, "Le centre fédéral de transfusion de l'AOF," *Médecine tropicale* 11, no. 6 (1951): 954–56. Thick smear leaves a drop of blood intact on a slide to detect more parasites quickly.

19. J. Linhard, G. Diebolt, and E. Ayité, "Particularités médicales des transfusions sous les tropiques," *Bulletin de la Société médicale d'Afrique noire de langue française* 18, no. 3 (1973): 293–97.

20. Almerindo Lessa et al., "Organisation de l'hématologie, l'hémothérapie, et la réanimation dans l'outre-mer portugais," *Ve Congrès International de Transfusion Sanguine* (Paris: Edition Septembre, 1955), 1057–61; Una Maclean, "Blood Donor Recruitment in Ibadan: The Record of One Year's Experience," *Journal of Tropical Medicine and Hygiene* 61, no. 12 (1958): 311–14; Central Pathology Laboratory [Dar es Salaam, Tanganyika], *Annual Report, 1960–61;* C.-S. Ronsée, "Anémies malariennes des enfants et transfusions sanguines, avec observations sur les groupes sanguins des Bakongo," *Mémoires, Institut Royal Colonial Belge, Section des sciences naturelles et médicales* 20, no. 2 (1952): 1–64; Kenya Medical Department, *Annual Report,* 1949–60.

21. Baba Sy, "Fonctionnement de la banque du sang de la Côte d'Ivoire," *Transfusion* (Paris) 3, no. 1 (1960): 47–51; N. O. Osamo, L. A. Okafor, and S. Enebe, "Blood Group Distribution in Nigerians in Relation to Foreign Nationalities: Haematological and Anthropological Implications," *East African Medical Journal* 59, no. 8 (1982): 546.

22. "Visit by Dr. K. L. G. Goldsmith to Tanzania, acting as a Temporary Consultant for the World Health Organization, 22nd—28th October and 28th October 1972," 3, AO658/2, International Federation of Red Cross and Red Crescent Societies, hereafter IFRC.

23. Ibid., 4–5.

24. "Rapport de Mission de Linhard."

25. G. Serafino, H. Tossou, and E. Goudote, "Réflexions sur cinq accidents transfusionnels par incompatibilité majeure," *Bulletin de la Société médicale d'Afrique noire de langue française* 6 (1961): 403–7; M. Sankalé, H. Ruscher, and Y. Touré, "Accidents et incidents de la transfusion sanguine et leur prévention dans un service de médecine à Dakar," *Bulletin de la Société médicale d'Afrique noire de langue française* 18, no. 3 (1973): 307–11.

26. *Rapport final du 1er Congrès africain de transfusion sanguine* (Yamoussoukro, Ivory Coast, 1977), 25.

27. *Uganda Red Cross Friend* (newsletter of Red Cross Youth), July 1973, 23–26, Uganda box 16723, IFRC.

28. Copy of registration form (received at IFRC, February 4, 1987, but probably dated late 1960s), 33, IFRC AO916, IFRC.

29. Aba S. David-West, "Blood Transfusion and Blood Bank Management in a Tropical Country," *Clinics in Haematology* 10, no. 3 (1981): 1013–28; J. O. Ndinya-Achola, H. Nsanzumuhire, and G. B. A. Okelo, "Some Possible Infectious Hazards Due to Blood Transfusion in Nairobi," *East African Medical Journal* 57, no. 1 (1980): 55–59.

30. Edington, "Blood Transfusion," 72–74.

31. L. J. Bruce-Chwatt, "Blood Transfusion and Tropical Disease," *Tropical Diseases Bulletin* 69, no. 9 (September 1972): 825–62; Linhard, Diebolt, and Ayité, "Particularités médicales," 293–97.

32. Bruce-Chwatt. "Blood Transfusion."

33. J. M. Goldsmid et al., "The Transmission of Blood Parasites via Blood Transfusion," *Central African Journal of Medicine* 20, no. 2 (1974): 23–30.

34. Edington, "Blood Transfusion"; Ndinya-Achola, Nsanzumuhire, and Okelo, "Infectious Hazards," 55–59; David-West, "Blood Transfusion."

35. This procedure was first attempted in Saigon in the 1950s. See L. J. Bruce-Chwatt, "Transfusion Malaria," *Bulletin of the World Health Organization* 50, nos. 3–4 (1974): 337–46; Sy, "Banque du sang," 145; David-West, "Blood Transfusion."

36. Goldsmid et al., "Blood Parasites;" P. R. Hira and S. F. Husein, "Some Transfusion-Induced Parasitic Infections in Zambia," *Journal of Hygiene, Epidemiology, Microbiology and Immunology* 23, no. 4 (1979): 436–44.

37. Baruch S. Blumberg, Harvey J. Alter, and Sam Visnich, "A 'New' Antigen in Leukemia Sera," *Journal of the American Medical Association* 191, no. 7 (February 15, 1965): 541–46. For early African studies, see Alison Parker, K. L. Muiruri, and J. K. Preston, "Hepatitis-Associated Antigen in Blood Donors in Kenya," *East African Medical Journal* 48, no. 9 (1971): 470–75; J. G. Cruickshank et al., "Australia Antigen—Rhodesia, 2: Surveys of Urban Blood Donors and Rural Populations," *Central African Journal of Medicine* 18, no. 6 (June 1972): 113–16. For a review of the early literature, see J. C. Zuzarte and E. G. Kasili, "Hepatitis B Antigen—A Review," *East African Medical Journal* 55, no. 8 (1978): 346–54.

38. Zuzarte and Kasili, "Hepatitis B Antigen."

39. R. W. Ryder et al., "Screening for Hepatitis B Virus Markers Is Not Justified in West African Transfusion Centres," *Lancet* 2, no. 8400 (August 25, 1984): 449–52.

40. M. Barbotin and J.-L. Oudart, "Reflexions sur l'incidence de la transfusion sanguine dans les hépatites en milieu africain," *Bulletin de la Société médicale d'Afrique noire de langue française* 18, no. 3 (1973): 298–302; G. Diebolt, J. Linhard, J. Seck, and V. Schwarzmann, "Fréquence de l'anti-gène Australia chez les donneurs du Centre National de Transfusion de Dakar," *Bulletin de la Société médicale d'Afrique noire de langue française* 16, no. 2 (1971): 231–37; Zuzarte and Kasili, "Hepatitis B Antigen."

41. See *Uganda Red Cross Friend*, July 1977, 23–26, .

42. David-West, "Blood Transfusion"; Linhard, Diebolt, and Ayité, "Particularités médicales."

43. *Rapport final du 1er Congrès africain de transfusion sanguine*. Other records are available at the WHO archives, Geneva.

44. Ibid., 25.

45. Ibid., 22.

46. World Health Organization and League of Red Cross Societies, "Report on the WHO/League of Red Cross Societies Workshop on the Development of Blood Transfusion Services, Nairobi, Kenya—12–17 November 1979," archives of the Kenya Red Cross Society, Nairobi. See also E. G. Kasili, "Blood Transfusion Services in a Developing Country: The Concept and Problems of Organization," *East African Medical Review* 58, no. 2 (1981): 81–83.

47. "Blood Bank and Kahn Test Report—1980," Kenya National Blood Transfusion Service, Nairobi, hereafter KNBTS.

48. League of Red Cross and Red Crescent Societies, *Optimal Use of Resources: African Workshop on Management of Blood Transfusion Services, December 2–11, 1985, Harare, Zimbabwe* (Geneva: LRCS, 1986), 36–7.

49. N. Clumeck et al., "Acquired Immune Deficiency Syndrome in Black Africans," *Lancet* 1, no. 8325 (March 19, 1983): 642; J. B. Brunet et al., "Acquired Immunodeficiency Syndrome in France," *Lancet* 1, no. 8326 (March 26, 1983): 700–701. On patients in Africa, see P. van de Perre et al., "Acquired Immunodeficiency Syndrome in Rwanda," *Lancet* 2, no. 8394 (July 14, 1984): 62–65; Peter Piot et al., "Acquired Immunodeficiency Syndrome in a Heterosexual Population in Zaire," *Lancet* 2, no. 8394 (July 14, 1984): 65–69.

50. See van de Perre et al., "Acquired Immunodeficiency Syndrome"; Mann et al., "Surveillance for AIDS." For an excellent summary, see T. Quinn, "AIDS in Africa."

51. Lieve Fransen, "Mission Report on Trip to Uganda—Kenya 6–15 May 87," Uganda Blood Transfusion Service records, Kampala. Personal papers of Prof. Esau Nzaro, Kampala.

52. J. W. Carswell, "HIV: Implications for Blood Transfusion and Banking in Africa," *Tropical Doctor* 20, no. 1 (January 1990): 42–43. See also Jonathan M. Mann, Daniel Tarantola, and Thomas W. Netter, *AIDS in the World* (Cambridge, MA: Harvard University Press, 1992), 436; and Jukka Koistinen, "Safe Blood: The WHO Sets Out Its Principles," *AIDS Analysis Africa* 2, no. 6 (November–December 1992): 4, 6.

53. W. Koinange, Director of Medical Services, to Director, National Public Health Laboratory, and all provincial medical officers, memorandum, June 10, 1985, KNBTS.

54. Gunnar Myllali, "Follow-up of the Development of the Blood Programme in Zimbabwe, 1982–1988," 12, FRCBS archives, Helsinki.

55. Mann et al., *AIDS in the World,* 437. The exception, Cameroon, reported that 90 percent of blood had been tested for HIV.

56. P. Ezeh, "Nigeria: Unscreened Blood Still Given to Patients." *New African* 255 (December 1988): 61; R. Okpara, "Transmission of HIV through Blood Transfusion," *African Health* 14, no. 5 (July 1992): 15–17.

57. H. Jäger et al., "Prevention of Transfusion-Associated HIV Transmission in Kinshasa, Zaire: HIV Screening Is Not Enough," *AIDS* 4, no. 6 (June 1990): 571–74; I. N'tita et al., "Risk of Transfusion-Associated HIV Transmission in Kinshasa Zaire," *AIDS* 5, no. 4 (1991): 437–39.

58. Projet Inter-Croix-Rouge de transfusion sanguine au Rwanda, "Rapport annuel du travail, 1985," Internationale Acties 7/24, Belgian Red Cross, Mechelen, hereafter BRC Mechelen.

59. Projet Inter-Croix-Rouge de transfusion sanguine au Rwanda, "Rapport annuel du travail, 1989," Internationale Acties 7/24, BRC Mechelen.

60. L. Stouffs to G. Hullebroeck, "Plan d'action de la lutte contre le SIDA," April 7, 1987, Internationale Acties 7/24, BRC Mechelen.

61. World Health Organization/League of Red Cross Societies, "Consensus Statement on Screening of Blood Donations for Infectious Agents Transmissible through Blood Transfusion, 30 January to 1 February 1990," WHO/LBS 91.1, WHO archives, Geneva.

62. For an example of a pilot program to respond to this situation, see G. Laleman et al., "Prevention of Blood-Borne HIV Transmission Using a Decentralized Approach in Shaba, Zaire," *AIDS* 6, no. 11 (November 1992): 1353–58.

63. Mann et al., *AIDS in the World.* Philippe Lepage and Philippe van de Perre found in studies done from 1984 to 1986 that "in children from Zaire, transfusion was

strikingly associated with HIV seropositivity. In hospital workers and children from Rwanda, blood transfusions were associated with positive HIV status." Lepage and van de Perre, "Nosocomial Transmission of HIV in Africa: What Tribute Is Paid to Contaminated Blood Transfusions and Medical Injections?" *Infection Control and Hospital Epidemiology* 9, no. 5 (May 1988): 200–203.

Chapter 7: African Blood Transfusion in the Context of Global Health

1. John Keegan, foreword to *A History of Military Medicine,* vol. 1, *From Ancient Times to the Middle Ages,* ed. Richard A. Gabriel and Karen S. Metz (New York: Greenwood, 1994), xi–xiii.

2. Philippe Lepage and Philippe van de Perre, "Nosocomial Transmission of HIV in Africa: What Tribute Is Paid to Contaminated Blood Transfusions and Medical Injections?" *Infection Control and Hospital Epidemiology* 9, no. 5 (May 1988): 200–203; William H. Schneider and Ernest Drucker, "Blood Transfusions in the Early Years of AIDS in Sub-Saharan Africa," *American Journal of Public Health* 96, no. 6 (June 2006): 984–94.

3. Still the most exhaustive analysis of predecessors of modern HIVs is Edward Hooper, *The River: A Journey to the Source of HIV and AIDS* (Boston: Little, Brown, 1999).

4. Mirko D. Grmek, *History of AIDS: Emergence and Origin of a Modern Pandemic,* trans. Russell C. Maulitz and Jacalyn Duffin (1989; repr., Princeton: Princeton University Press, 1990), 161.

5. Peter Piot and Michael Bartos, "The Epidemiology of HIV and AIDS," in *AIDS in Africa,* 2nd ed., ed. Max Essex et al. (New York: Kluwer Academic Publishers, 2002), 200–17. More recent studies differ slightly in rates for each method of transmission, but agree in the order of magnitude difference between them. See Rebecca F. Baggaley, Marie-Claude Boily, Richard G. White, and Michel Alary, "Risk of HIV-1 Transmission for Parenteral Exposure and Blood Transfusion: A Systematic Review and Meta-analysis," *AIDS* (London) 20, no. 6 (April 2006): 805–12; and Marie-Claude Boily et al., "Heterosexual risk of HIV-1 Infection per Sexual Act: Systematic Review and Meta-analysis of Observational Studies," *Lancet Infectious Diseases* 9, no. 2 (February 2009): 118–29.

6. Preston A. Marx, Cristian Apetrei, Ernest Drucker, "AIDS as a Zoonosis? Confusion over the Origin of the Virus and the Origin of the Epidemics," *Journal of Medical Primatology* 33, no. 5–6 (October 2004): 220–26. A zoonosis is a disease normally communicable from animals to humans.

7. "Simian Viruses and Emerging Diseases in Humans," symposium, June 9–10, 2010, Paris. Presentations and discussions available at http://www.simianviruses.univ-paris-diderot.fr/SympoJune2010/ (last accessed March 22, 2013).

8. Preston A. Marx, Phillip G. Alcabes, and Ernest Drucker, "Serial Human Passage of Simian Immunodeficiency Virus by Unsterile Injections and the Emergence of Epidemic Human Immunodeficiency Virus in Africa," *Philosophical Transactions of the Royal Society of London, B: Biological Sciences* 356, no. 1410 (2001): 911–20.

9. Michael Worobey et al., "Direct Evidence of Extensive Diversity of HIV-1 in Kinshasa by 1960," *Nature* 455, no. 7213 (2008): 661–64; Joel O. Wertheim and Michael Worobey, "Dating the Age of the SIV Lineages That Gave Rise to HIV-1 and HIV-2," *PLoS Computational Biology* 5, no. 5 (October 2009), e1000377.

10. Philippe Lemey et al., "Tracing the Origin and History of the HIV-2 Epidemic," *Proceedings of the National Academy of Sciences of the United States of America* 100, no. 11 (May 2003): 6588–92.

11. For central Africa, see Jacques Pepin and Annie-Claude Labbé, "Noble Goals, Unforeseen Consequences: Control of Tropical Diseases in Colonial Central Africa and the Iatrogenic Transmission of Blood-Borne Viruses," *Tropical Medicine and International Health* 13 (June 2008): 744–53. For smallpox, see William H. Schneider, "Smallpox in Africa during Colonial Rule," *Medical History* 53 (April 2009): 193–227.

12. See, for example, Anne-Emmanuel Birn, "Gates's Grandest Challenge: Transcending Technology as Public Health Ideology," *Lancet* 366, no. 9484 (August 6, 2005): 514–19. For a more general critique of all outside assistance, see Dambisa Moyo, *Dead Aid: Why Aid Is Not Working and How There Is Another Way for Africa* (New York: Farrar, Straus and Giroux, 2009).

13. WHO and International Federation of Red Cross and Red Crescent Societies, *Towards 100% Voluntary Blood Donation: A Global Framework for Action* (Geneva: World Health Organization, 2010), accessed November 12, 2011, http://www.who.int/bloodsafety/publications/9789241599696_eng.pdf.

BIBLIOGRAPHY

Archival Sources

Archives africaines, Ministère des Affaires étrangères et de commerce extérieur, Brussels
Archives of the Belgian Red Cross, Belgian State Archives, Brussels
Archives nationales du Sénégal
Belgian Red Cross, Mechelen
British Red Cross Museum and Archives, London
Centre national de transfusion sanguine, Dakar, Senegal (uncataloged archives)
Croix-Rouge du Congo, Archives of Belgian Red Cross, Archives royales de Belgique, Brussels
Finnish Red Cross Blood Transfusion Service, Helsinki
Institut de médecine tropicale du Service de santé des armées, Marseille
Institut Pasteur, Paris
International Federation of Red Cross and Red Crescent Societies, Geneva
Kenya National Archives, Nairobi
Kenya National Blood Transfusion Service, Nairobi (uncataloged archives)
Musée international de la Croix-Rouge, Geneva
Oxford Rhodes Library, Oxford University
Senegal, National Archives, Dakar
Swiss Red Cross archives, Bern
Uganda Blood Transfusion Service, Kampala (uncataloged archives)

Journals and Published Reports

Afrique Occidentale Française. Service de santé. *Rapport annuel.* 1939, 1940.

British Red Cross Society. *Annual Report.* 1932, 1939, 1942–45.

———. *British Empire Red Cross Conference, 1930.* London: British Red Cross Society, 1930.

———. *British Red Cross Quarterly Review.* 1939–45.

———. *Report of the Overseas Branches.* 1950.

Fédération internationale des organisations de donneurs de sang. *Le don universel du sang,* 1967–80.

Fifth International Congress of Blood Transfusion, Paris 1954. Paris: Édition Septembre, 1955.

Gold Coast Colony. *Departmental Reports,* 1935–36.

———. *Medical Research Institute.* Report 1954–57.

Institut Pasteur (Cameroun). *Rapport annuel.* 1960–69.

Institut Pasteur (Senegal). "Rapport sur le fonctionnement technique de l'Institut Pasteur de Dakar." 1945, 1948.

Kenya Colony and Protectorate. *Medical Department Annual Report.* 1957, 1959.

———. *Medical Research Laboratory Annual Report*, 1933–37.
———. *Ministry of Health Annual Report*. 1959, 1962.
Kenya. Medical Department. *Annual Report*. 1949–60, 1966.
Kenya, Republic of. Ministry of Health. *Annual Report*, 1965.
League of Red Cross and Red Crescent Societies. *Optimal Use of Resources: African Workshop on Management of Blood Transfusion Services, December 2–11, 1985, Harare, Zimbabwe*. Geneva: LRCS, 1986.
Rapport final du 1er Congrès africain de transfusion sanguine. Yamoussoukro, Ivory Coast, 1977.
République du Sénégal. Ministère de la santé publique. *Activité du service*. 1961, 1969.
———. *Statistiques sanitaires*. 1963–69, 1976, 1977.
Republic of Zambia. *Ministry of Health Annual Report*. 1965, 1966, 1972.
Service de l'hygiène et laboratoires bactériologiques (Léopoldville). *Rapport annuel*. 1937.
Service médical, Union minière du Haut Katanga. *Rapport annuel*. 1929, 1932, 1939.
Sierra Leone. *Annual Report, Medical and Health Services*. 1949, 1950.
———. *Annual Report of the Medical and Sanitary Department*. 1936.
———. *Report on the Medical and Health Service*. 1959, 1970.
Tanganyika. Central Pathology Laboratory (Dar es Salaam). *Annual Report*. 1960–61.
Tanganyika Territory. *Annual Medical and Sanitary Report*. 1932.
———. *Annual Report of the Medical Department [Laboratory]*. 1947, 1954, 1957, 1958.
Uganda. Ministry of Health. *Annual Report*, 1961–63.
Uganda Protectorate. *Annual Medical and Sanitary Report*. 1931, 1957.
Uganda Red Cross Society. *Annual Report*. 1982–86.

Newspapers

Daily Nation (Nairobi), 1969.
East African Standard (Nairobi), 1969.
Sunday Nation (Nairobi), 1971.
Voix du congolais (Léopoldville), 1954.

Secondary Sources

Abdalla, F., O. W. Mwanda, and F. Rana. "Comparing Walk-in and Call-Responsive Donors in a National and a Private Hospital in Nairobi." *East African Medical Journal* 82, no. 10 (October 2005): 531–35.
Abdallah, Saade, Rebekah Heinzen, and Gilbert Burnham. "Immediate and Long-Term Assistance Following the Bombing of the U.S. Embassies in Kenya and Tanzania." *Disasters* 31, no. 4 (December 2007): 417–34.
Adant, M. "L'organisation de la transfusion sanguine en Belgique. In *Fifth International Congress of Blood Transfusion*. Paris 1954, 999. Paris: Edition Septembre, 1955
Addo-Yobo, E. O., and H. Lovel. "How Well Are Hospitals Preventing Iatrogenic HIV? A Study of the Appropriateness of Blood Transfusions in Three Hospitals in the Ashanti Region, Ghana." *Tropical Doctor* 21, no. 4 (1991): 162–64.
Águas, Ricardo, Lisa J. White, Robert W. Snow, and M. Gabriela M. Gomes. "Prospects for Malaria Eradication in Sub-Saharan Africa." *PLoS ONE* 12, no. 3 (March

2008): e1767. Accessed September 29, 2011. http://www.ncbi.nlm.nih.gov/pmc/articles/PMC2262141/.

Akué, B. A. [Title unknown.] Medical thesis, Université de Dakar, [1981?].

Allain, Jean-Pierre, Shirley Owusu-Ofori, and Imelda Bates. "Blood Transfusion in Sub-Saharan Africa." *Transfusion Alternatives in Transfusion Medicine* 6, no. 1 (March 2004): 16–23.

Alnwick, D. J., M. R. Stirling, and G. Kyeyune. "Population Access to Hospitals, Health Centres and Dispensary/Maternity Units in Uganda, 1980." In Dodge and Wiebe, *Crisis in Uganda,* 193–200.

Alsted, Gunnar. "Red Cross Participation in Blood Transfusion." In *Fifth International Congress of Blood Transfusion, Paris 1954,* 185–88. Paris: Edition Septembre, 1955.

Alter, Harvey J., and Harvey G. Klein. "The Hazards of Blood Transfusion in Historical Perspective." *Blood* 112, no. 7 (October 2008): 2617–26.

American Association of Blood Banks. The 2005 Nationwide Blood Collection and Utilization Survey Report. Bethesda, MD: AABB, 2005. Accessed on March 13, 2013. http://www.aabb.org/programs/biovigilance/nbcus/Documents/05nbcusrpt.pdf.

Anangwe, Alfred. "Health Sector Reforms in Kenya: User Fees." In *Governing Health Systems in Africa,* edited by Martyn Sama and Vinh-Kim Nguyen, 44–59. Dakar: Codesria, 2008.

André, J., J. Burke, J. Vuylsteke, and H. van Balen. "Evolution of Health Services." In Janssens, Kivits, and Vuylsteke, *Health in Central Africa,* 89–158.

Anorlu, R. I., C. O. Orakwe, O. O. Abudu, and A. S. Akanmu. "Uses and Misuse of Blood Transfusion in Obstetrics in Lagos, Nigeria." *West African Journal of Medicine* 22, no. 2 (June 2003): 124–27.

Apetrei, C., A. Kaur, N. W. Lerche, M. Metzger, I. Pandrea, J. Hardcastle, S. Falkenstein, et al. "Molecular Epidemiology of Simian Immunodeficiency Virus SIVsm in U.S. Primate Centers Unravels the Origin of SIVmac and SIVstm." *Journal of Virology* 79, no. 14 (July 2005): 8991–9005.

Atangana, Martin-René. *French Investment in Colonial Cameroon: The FIDES Era (1946–1957).* New York: Peter Lang, 2009.

Ayité, Etienne. "La transfusion sanguine en Afrique noire de langue française." Medical thesis, Faculté de médecine et de pharmacie, Dakar, 1974.

Ayité, Etienne G., G. Diebolt, and J. Linhard. "La transfusion sanguine au Sénégal." *Bulletin de la Société médicale d'Afrique noire de langue française* 18, no. 3 (1973): 289–92.

Bado, Jean-Paul. *Les conquêtes de la médecine moderne en Afrique.* Paris: Karthala, 2006.

Barbotin, M., and J.-L. Oudart. "Reflexions sur l'incidence de la transfusion sanguine dans les hépatites en milieu africain." *Bulletin de la Société médicale d'Afrique noire de langue française* 18, no. 3 (1973): 298–302.

Barradas, F. R., T. Schwalbach, and A. Novoa. "Blood and Blood Products Usage in Maputo." *Central African Journal of Medicine* 40, no. 3 (1994): 56–60.

Bates, I., G. Manyasi, and A. Medina Lara. "Reducing Replacement Donors in Sub-Saharan Africa: Challenges and Affordability." *Transfusion Medicine* 17, no. 6 (December 2007): 434–42.

Bates, Robert H. *When Things Fell Apart: State Failure in Late-Century Africa.* Cambridge: Cambridge University Press, 2008.

Benhamou, Edmond. "Notes pour servir à l'histoire de la transfusion sanguine dans l'armée française de 1942 à 1945 à partir de l'Afrique du Nord." *Revue des Corps de santé des Armées Terre, Mer, Air* 7, special issue (November 1966): 859–62.

Bird, G. W. "Percy Lane Oliver, OBE (1878–1944): Founder of the First Voluntary Blood Donor Panel." *Transfusion Medicine* 2, no. 2 (June 1992): 159–60.

Birn, Anne-Emanuelle. "Gates's Grandest Challenge: Transcending Technology as Public Health Ideology." *Lancet* 366, no. 9484 (August 6, 2005): 514–19.

Blanchard, Maurice. "L'École de médecine de l'Afrique Occidentale Française de sa fondation à l'année 1934." *Annales de médecine et de pharmacie coloniale* 33 (1934): 90–111.

"Blood Donation—The Jomo Kenyatta Blood Donation Week." Editorial. *Medicus* 3, no. 11 (November 1984): 1.

Blumberg, Baruch S. *Hepatitis B: The Hunt for a Killer Virus.* Princeton: Princeton University Press, 2003.

Blumberg, Baruch S., Harvey J. Alter, and Sam Visnich. "A 'New' Antigen in Leukemia Sera." *Journal of the American Medical Association* 191, no. 7 (February 15, 1965): 541–46.

Bouyer, C. "Perfusions et transfusions par la veine sous-clavière chez les nourissons et les enfants." *Annales de la Société belge de médecine tropicale* 45, no. 1 (1965): 39–48.

Bruce-Chwatt, L. J. "Blood Transfusion and Tropical Disease." *Tropical Diseases Bulletin* 69, no. 9 (September 1972): 825–62.

———. "Transfusion Malaria." *Bulletin of the World Health Organization* 50, nos. 3–4 (1974): 337–46.

Brunet, J. B., E. Bouvet, J. Leibowitch, J. Chaperon, C. Mayaud, J. C. Gluckman, O. Picard, et al. "Acquired Immuniodeficiency Syndrome in France." *Lancet* 1, no. 8326 (March 26, 1983): 700–701.

Brunet-Jailly, Joseph. "La santé dans quelques pays d'Afrique de l'Ouest après quinze ans d'ajustement." In *Crise et population en Afrique: Crises économiques, politiques d'ajustement et dynamiques démographiques,* edited by Jean Coussy and Jacques Vallin, 233–71. Paris: Centre français sur la population et développement, 1996. Accessed June 23, 2012. http://www.ceped.org/cdrom/integral_publication_1988_2002/etudes/pdf/etudes_cpd_13.pdf.

Brunschwig, Henri. *Noirs et blancs dans l'Afrique noire française, ou, Comment le colonisé devient colonisateur, 1870–1914.* Paris: Flammarion, 1983.

Buckley, Thomas C.T., and Alma Gottlieb. *Blood Magic: The Anthropology of Menstruation* Berkeley: University of California Press, 1988

Burke, J. "Historique de la lutte contre la maladie du sommeil au Congo." *Annales de la Société belge de médecine tropicale* 51, no. 4 (1971): 465–82.

Carman, John A. *A Medical History of the Colony and Protectorate of Kenya: A Personal Memoir.* London: Rex Collings, 1976.

Carswell, J. W. "HIV: Implications for Blood Transfusion and Banking in Africa." *Tropical Doctor* 20, no. 1 (January 1990): 42–43.

Casteret, Ann-Marie. *L'affaire du sang.* Paris: La Découverte, 2002.

Chauveau, Sophie. "De la transfusion à l'industrie: Une histoire des produits sanguins en France (1950–fin des années 1970)." *Entreprises et histoire* 2, no. 36 (October 2004): 103–19.

———. "Du don à l'industrie: La transfusion sanguine en France depuis les années 1940." *Terrain: Revue d'ethnologie de l'Europe,* no. 56 (March 2011): 74–89.

Chippaux, Claude. "Le service de santé des troupes de marine." *Médecine tropicale* 40, no. 6 (1980): 605–30.

Chitnis, Amit, Diana Rawls, and Jim Moore. "Origin of HIV Type 1 in Colonial French Equatorial Africa?" *AIDS Research and Human Retroviruses* 16, no. 1 (January 2000): 5–8.

Clumeck, N., F. Mascart-Lemone, J. de Maubeuge, D. Brenez, and L. Marcelis. "Acquired Immune Deficiency Syndrome in Black Africans." *Lancet* 1, no. 8325 (March 19, 1983): 642.

Collins, David, Grace Njeru, Julius Meme, and William Newbrander. "Hospital Autonomy: The Experience of Kenyatta National Hospital." *International Journal of Health Planning and Management* 14, no. 2 (1999): 129–53.

Conombo, Joseph. "Croix-Rouge voltaïque." *Don universel du sang* 22 (1972): 19–20.

Coulibaly, Seydou, and Moussa Keita. *Les comptes nationaux de la santé au Mali, 1988–1991.* Bamako: Institut National de Recherche en Santé Publique, 1993.

Crile, George. *George Crile: An Autobiography.* Edited by Grace Crile. 2 vols. Philadelphia: Lippincott, 1947.

Crozier, Anna. *Practising Colonial Medicine: The Colonial Medical Service in British East Africa.* New York: I. B. Tauris, 2007.

Cruickshank, J. G., R. Swanepoel, R. F. Lowe, T. Robertson, and H. Moore. "Australia Antigen—Rhodesia, 2: Survey of Urban Blood Donors and Rural Populations." *Central African Journal of Medicine* 18, no. 6 (June 1972): 113–16.

David-West, Aba S. "Blood Transfusion and Blood Bank Management in a Tropical Country." *Clinics in Haematology* 10, no. 3 (October 1981): 1013–28.

de Smet, M. P. "Traitement d'attaque des anémies-oedèmes graves par transfusions fractionnées chez des enfants sevrés." *Annales de la Société belge de médecine tropicale* 34, no. 2 (1954): 155–69.

Dewhurst, K. S. "Observations on East African Blood Donors." *East African Medical Journal* 22 (1945): 276–78.

Dhingra, Neelam. "Making Safe Blood Available in Africa." Statement before Committee on International Relations, Subcommittee on Africa, Global Human Rights and International Operations, U.S. House of Representatives, June 27, 2006. http://www.who.int/bloodsafety/makingsafebloodavailableinafricastatement.pdf. Accessed April 30, 2010.

Diebolt, G., J. Linhard, J. Seck, and V. Schwarzmann. "Fréquence de l'anti-gène Australia chez les donneurs du Centre national de transfusion de Dakar." *Bulletin de la Société médicale d'Afrique noire de langue française* 16, no. 2 (1971): 231–37.

Dodge, Cole P., and Paul D. Wiebe, eds. *Crisis in Uganda: The Breakdown of Health Services.* Oxford: Pergamon Press, 1985.

Dodsworth, H., and H. H. Gunson. "Fifty Years of Blood Transfusion." *Transfusion Medicine* 6, suppl. 1 (1996): 1–88.

Duboccage, Antoine. "La Fondation médicale de l'Université de Louvain au Congo." *Lovania* (Elisabethville) 3 (1944): 6–10.

———. "Notes cliniques du service gynécologique de la Fomulac à Kisantu." *Annales de la Société belge de médecine tropical* 14 (1934): 421–33.

Dubois, Albert. *La Croix-Rouge au Congo. Classe des sciences naturelles et médicales,* new series 18–2. Brussels: Académie royale des sciences d'outre-mer, 1969, 1–61.

———. "Jeanne Legrand (1916–60)." *Annales de la Société belge de médecine tropicale* 40 (1960): 711.

Dubois, Albert, and Albert Duren. "Soixante ans d'organisation médicale au Congo belge." *Annales de la Société belge de médecine tropicale* 27 (1947): 1–36.

Dulles, Foster Rhea. *The American Red Cross: A History.* New York: Harper, 1950.

Echenberg, Myron. *Black Death, White Medicine: Bubonic Plague and the Politics of Public Health in Colonial Senegal, 1914–1945.* Portsmouth, NH: Heinemann, 2002.

"L'école de médecine indigène de l'Afrique occidentale française." *Bulletin d'informations et de renseignements* (AOF) 199 (August 15, 1938): 303–6.

Edington, G. M. "Some Observations on Blood Transfusion in the Gold Coast." *West African Medical Journal* 5, no. 2 (1956): 71–75.

Elmendorf, A. Edward. *Structural Adjustment and Health in Africa in the 1980s.* Washington, DC: American Public Health Association, 1993.

Emeribe, A. O., A. O. Ejele, E. E. Attai, and E. A. Usanga. "Blood Donations and Patterns of Use in Southeastern Nigeria." *Transfusion* 33, no. 4 (1993): 330–32.

Enosolease, M. E., C. O. Imarengiaye, and O. A. Awodu. "Donor Blood Procurement and Utilisation at the University of Benin Teaching Hospital, Benin City." *African Journal of Reproductive Health* 8, no. 2 (August 2004): 59–63.

Établissement français du sang. "Donner son sang en France, quatrième edition, Juillet 2007." Paris: Établissement français du sang. Accessed August 26, 2011. http://www.cerphi.org/wp-content/uploads/2011/05/Don-sang-2007.pdf.

Ezeh, P. "Nigeria: Unscreened Blood Still Given to Patients." *New African* 255 (December 1988): 61.

Farr, A. D. "The First Human Blood Transfusion." *Medical History* 24, no. 2 (1980): 143–62.

Fassin, Didier. "Le domaine privé de la santé publique: Pouvoir, politique et sida au Congo." *Annales: Histoire, sciences sociales* 49, no. 4 (1994): 745–75.

Fleming, Alan F. "HIV and Blood Transfusion in Sub-Saharan Africa." *Transfusion Science* 18, no. 2 (June 1997): 167–79.

Gasson, S. G. "The First Blood Transfusion in Southern Rhodesia." *Central African Journal of Medicine* 6 (November 1960): 490.

Gear, J. H. S. "Blood Transfusion in Johannesburg, Pre–South African Blood Transfusion Service." *Adler Museum Bulletin* 14, no. 1 (June 1988): 3–6.

Germond. "Statistiques des cas de pneumonie traités par transfusion de sang de convalescents." *Bulletin médical du Katanga* 1 (1924): 243.

Ghose, R. "History of Blood Transfusion in Ethiopia." *Ethiopian Medical Journal* 31, no. 4 (April 1963): 208–12.

"Gift of Life, The." *Spotlight* (Geneva: League of Red Cross and Red Crescent Societies) 8 (1990).

Giles-Vernick, Tamara, Didier Gondola, Guillaume Lachenal, and William H. Schneider. "The Emergence of HIV/AIDS: An Historical Review." *Journal of African History* 54, no. 1 (2013): 1–20.

Goldsmid, J. M., R. F. Lowe, S. Rogers, L. Cranston, and T. Robertson. "The Transmission of Blood Parasites via Blood Transfusion." *Central African Journal of Medicine* 20, no. 2 (1974): 23–30.

Greenberg, Alan E., Phuc Nguyen-Dinh, Jonathan M. Mann, Ndoko Kabote, Robert L. Colebunders, Henry Francis, Thomas C. Quinn, et al. "The Association between Malaria, Blood Transfusions, and HIV Seropositivity in a Pediatric Population in Kinshasa, Zaire." *Journal of the American Medical Association* 259, no. 4 (1988): 545–49.

Greenwalt, Tibor. *History of the International Society of Blood Transfusion, 1935–1995.* Groningen: Stichting Transfusion Today Foundation, 2000.

Grmek, Mirko D. *History of AIDS: Emergence and Origin of a Modern Pandemic.* Translated by Russell C. Maulitz and Jacalyn Duffin. 1989. Reprint, Princeton: Princeton University Press, 1990.

Harrison, Mark, Margaret Jones, and Helen Sweet, eds. *From Western Medicine to Global Medicine: The Hospital beyond the West.* New Delhi: Orient Blackswan, 2009.

Hennebert, Paul Norbert. "La médicine à Lovanium." Accessed September 25, 2006. http://www.md.ucl.ac.be/histoire/livre/lovan.pdf.

Hermitte, Marie-Angèle. *Le sang et le droit: Essai sur la transfusion sanguine.* Paris: Éditions de Seuil, 1996.

Hira, P. R., and S. F. Husein. "Some Transfusion-Induced Parasitic Infections in Zambia." *Journal of Hygiene, Epidemiology, Microbiology and Immunology* 23, no. 4 (1979): 436–44.

Holmberg, Jerry A., Sridhar Basavaraju, Bakary Drammeh, and Michael Qualls. "Progress toward Strengthening Blood Transfusion Services—14 Countries, 2003–2007." *Morbidity and Mortality Weekly Report* 57, no. 47 (November 28, 2008): 1273–77.

Holzer, Benedik R., Matthias Egger, Thomas Teuscher, Stefan Koch, Dominik M. Mboya, and George Davey Smith. "Childhood Anemia in Africa: To Transfuse or Not Transfuse?" *Acta tropica* 55, no. 1 (October 1993): 47–51.

Hooper, Edward. *The River: A Journey to the Source of HIV and AIDS.* Boston: Little, Brown, 1999.

Hunt, Nancy Rose. *A Colonial Lexicon of Birth Ritual, Medicalization, and Mobility in the Congo.* Durham, NC: Duke University Press, 1999.

———. "Normality and a 'Disease of Civilization': Eclampsia and Race in the Congo and the U.S. South." In *Discovering Normality in Health and the Reproductive Body,* proceedings of a workshop held at the Program of African Studies, Northwestern University, March 9–10, 2001, edited by Caroline H. Bledsoe, 125–35. Evanston: Program on African Studies, 2002.

Iliffe, John. *The African AIDS Epidemic: A History.* Athens: Ohio University Press, 2006.

———. *East African Doctors: A History of the Modern Profession.* Cambridge: Cambridge University Press, 1998.

"Injecting Blood Royal, or, Phlebotomy at St. Cloud." London: W. Holland, June 1804. In *Images from the History of Medicine (NLM).* Accessed April 24, 2012. http://ihm.nlm.nih.gov/images/A21765.

Jäger, H., B. N'Galy, J. Perriens, K. Nseka, F. Davachi, C. M. Kabeya, G. Rauhaus, et al. "Prevention of Transfusion-Associated HIV Transmission in Kinshasa, Zaire: HIV Screening Is Not Enough." *AIDS* 4, no. 6 (June 1990): 571–74.

Janssens, P. G. "Eugène Jamot et Émile Lejeune: Pages d'histoire." *Annales de la Société belge de médecine tropicale* 75, no. 1 (March 1995): 10–11.

Janssens, P. G., M. Kivits, and J. Vuylsteke, eds. *Health in Central Africa since 1885: Past, Present and Future.* Brussels: King Baudouin Foundation, 1997.

Jayaraman, S., Z. Chalabi, P. Perel, C. Guerriero, and I. Roberts. "The Risk of Transfusion-Transmitted Infections in Sub-Saharan Africa." *Transfusion* 50, no. 2 (February 2010): 433–42.

Johnson, Ryan. "Historiography of Medicine in British Colonial Africa." *Global South* 6, no. 3 (2010): 20–28.

Kamdem, Sandrine Simeu. "Evolution de la pratique de la transfusion sanguine dans 3 hôpitaux du Cameroun." Medical thesis, University of Yaoundé, 2010.

Kasili, E. G. "Blood Transfusion Services in a Developing Country: The Concept and Problems of Organization." *East African Medical Review* 58, no. 2 (1981): 81–83.

Keegan, John. Foreword to *A History of Military Medicine:* Volume 1: *From Ancient Times to the Middle Ages,* edited by Richard A. Gabriel and Karen S. Metz, xi–xiii. New York: Greenwood, 1992.

Keita, Maghan. *A Political Economy of Health Care in Senegal.* Leiden, Brill: 2007.

Kerouedan, D., W. Bontez, A. Bondurand, S. Abisse, and S. Konate. "Réflexions sur la transfusion sanguine en Afrique au temps de l'épidémie de sida: État des lieux et perspectives en Côte d'Ivoire." *Santé* 4, no. 1 (1994): 37–42.

Keynes, Geoffrey. *Blood Transfusion.* London: Henry Frowde and Hodder and Stoughton, 1922.

Kifuani, Nseka. "La situation actuelle de la transfusion dans notre pays." *Hôpital africain* 15 (1983): 18.

Kilgore, A. R., and J. H. Liu. "Isoagglutination Tests of Chinese Bloods for Transfusion Compatibility." *China Medical Journal* 3 (1918): 21–25.

Koistinen, Jukka. "Safe Blood: The WHO Sets Out Its Principles." *AIDS Analysis Africa* 2, no. 6 (November–December 1992): 4, 6.

Kok, L. "Les transfusions de sang en milieu indigène." Medical thesis, Prince Leopold Institute of Tropical Medicine, Antwerp, 1950.

Lachenal, Guillaume, Preston Marx, William H. Schneider, Ernest Drucker, and François Simon. "Simian Viruses and Emerging Diseases in Human Beings." *Lancet* 376, no. 9756 (December 4, 2010): 1901–2.

Lackritz, Eve M., T. K. Ruebush 2nd, J. R. Zucker, J. E. Adungosi, J. B. Were, and C. C. Campbell. "Blood Transfusion Practices and Blood-Banking Services in a Kenyan Hospital." *AIDS* 7, no. 7 (1993): 995–99.

Laleman, G., K. Magazani, J. H. Perriens, N. Badibanga, N. Kapila, M. Konde, U. Selemani, and P. Piot. "Prevention of Blood-Borne HIV Transmission Using a Decentralized Approach in Shaba, Zaire." *AIDS* 6, no. 11 (November 1992): 1353–58.

Lambillon, Joseph. "Contribution à l'étude du problème obstétrical chez l'autochtone du Congo belge." *Annales de la Société belge de médecine tropicale* 30, no. 5 (1950): 987–1123.

———. "Fomulac Katana: Deuxième centre de la Fondation médicale de l'Université de Louvain au Congo." *Lovania* 5 (1944): 5–10.

Lambillon, Joseph, and N. Denisoff. "Étude de l'organisation d'un service de transfusions sanguines dans un centre hospitalier d'Afrique." *Annales de la Société belge de médecine tropicale* 20 (1940): 279–85.

Langeron, Jean. "Étude sur la sérotherapie de la pneumonie: Valeurs thérapeutiques des sérums des convalescents." *Bulletin médical du Katanga* 1 (1924): 231.

Lapeyssonnie, Léon. *La médecine coloniale: Mythes et réalités*. Paris: Seghers, 1988.

Lederer, Susan E. *Flesh and Blood: Organ Transplantation and Blood Transfusion in Twentieth-Century America*. New York: Oxford University Press, 2008.

Leikola, Juhani. "How Much Blood for the World?" *Vox sanguinis* 54, no. 1 (1988): 2.

Leikola, Juhani, and W. G. van Aken. "The Story of *Vox sanguinis*." *Vox sanguinis* 100, no. 1 (2011): 2–9.

Lejeune, Émile. "Transfusion sanguine après hémoglobinurie grave." *Annales de la Société belge de médecine tropicale* 1 (1921): 299–300.

Lemey, Philippe, Oliver G. Pybus, Bin Wang, Nitin K. Saksena, Marco Salemi, and Anne-Mieke Vandamme. "Tracing the Origin and History of the HIV-2 Epidemic." *Proceedings of the National Academy of Sciences of the United States of America* 100, no. 11 (May 2003): 6588–92.

Lepage, Philippe, and Philippe van de Perre. "Nosocomial Transmission of HIV in Africa: What Tribute Is Paid to Contaminated Blood Transfusions and Medical Injections?" *Infection Control and Hospital Epidemiology* 9, no. 5 (May 1988): 200–203.

Lessa, Almerindo, L. Mayor, P. de Figueiredo, and A. Rebeko. "Organisation de l'hématologie, l'hémothérapie, et la réanimation dans l'outre-mer portugais." In *Ve Congrès international de transfusion sanguine* (Paris: Édition Septembre, 1955).

Levene, Cyril. "Brief History of the Development of the Transfusion Service." *How to Recruit Voluntary Donors in the Third World?* Geneva: LRCS, 1984.

Lhoiry, J. "L'utilisation de la transfusion sanguine outre-mer dans les centres chirurgicaux secondaires." *Médecine tropicale* 14, no. 5 (1954): 569–79.

Lindboom, G. A. "The Story of a Blood Transfusion to a Pope." *Journal of the History of Medicine and Allied Sciences* 9, no. 4 (1954): 455–59.

Linhard, Jacques. "Le centre fédéral de transfusion de l'AOF." *Médecine tropicale* 11, no. 6 (1951): 951–57.

Linhard, Jacques, G. Diebolt, and E. Ayité. "Particularités médicales des transfusions sous les tropiques." *Bulletin de la Société médicale d'Afrique noire de langue française* 18, no. 3 (1973): 293–97.

Lodewyck, A. "Note sur la transfusion sanguine chez les nourrissons et les enfants." *Recueil de travaux de sciences médicales au Congo belge* 2 (1944): 157–61.

Lucas, Adetokunbo A. "What We Inherited: An Evaluation of What Was Left Behind at Independence and Its Effects on Health and Medicine Subsequently." In *Health in Tropical Africa during the Colonial Period*, edited by E. E. Sabben-Clare, David J. Bradley, and Kenneth Kirkwood, 239–48. Oxford: Clarendon Press, 1980.

Lyons, Maryinez. "Public Health in Colonial Africa: The Belgian Congo." In *The History of Public Health and the Modern State*, edited by Dorothy Porter, 356–84. Amsterdam: Editions Rodopi, 1994.

MacDonald, G. "Theory of the Eradication of Malaria." *Bulletin of the World Health Organization* 15, nos. 3–5 (1956): 369–87.
Maclean, Una. "Blood Donor Recruitment in Ibadan: The Record of One Year's Experience." *Journal of Tropical Medicine and Hygiene* 61, no. 12 (1958): 311–14.
———. "Blood Donors for Nigeria." *Community Development Bulletin* 11 (1960): 26–31.
Maluf, N. S. R. "History of Blood Transfusion: The Use of Blood from Antiquity through the Eighteenth Century." *Journal of the History of Medicine and Allied Sciences* 9, no. 1 (1954): 59–107.
Mann, Jonathan M., H. Francis, T. Quinn, P. K. Asila, N. Bosenge, N. Nzilambi, K. Bila, et al. "Surveillance for AIDS in a Central African City: Kinshasa, Zaire." *Journal of the American Medical Association* 255, no. 23 (June 20, 1986): 3255–59.
Mann, Jonathan M., Daniel Tarantola, and Thomas W. Netter. *AIDS in the World.* Cambridge, MA: Harvard University Press, 1992.
Marx, Preston A., Phillip G. Alcabes, and Ernest Drucker. "Serial Human Passage of Simian Immunodeficiency Virus by Unsterile Injections and the Emergence of Epidemic Human Immunodeficiency Virus in Africa." *Philosophical Transactions of the Royal Society of London, B: Biological Sciences* 356, no. 1410 (2001): 911–20.
Marx, Preston A., Cristian Apetrei, and Ernest Drucker. "AIDS as a Zoonosis? Confusion over the Origin of the Virus and the Origin of the Epidemics." *Journal of Medical Primatology* 33, nos. 5–6 (October 2004): 220–26.
Massenet, D., G. Tesfaye, and B. Dandera. "La transfusion sanguine en Ethiopie." *Médecine tropicale* 58, no. 3 (1998): 307–8.
Mbanya, Dora, Fidele Binam, and Lazare Kaptué. "Transfusion Outcome in a Resource-Limited Setting of Cameroon: A Five-Year Evaluation." *International Journal of Infectious Diseases* 5, no. 2 (2001): 70–73.
Mburu, F. M. "Socio-political Imperatives in the History of Health Development in Kenya." *Social Science and Medicine, A: Medical Sociology* 15, no. 5 (1981): 521–27.
Mignonsin, D., S. Abissey, B. Vilasco, M. Kane, and A. Bondurand. "Transfusion sanguine en Côte d'Ivoire: Perspectives d'avenir." *Médecine d'Afrique noire* 38, no. 11 (1991): 723–31.
Moody, Elizabeth. "Blood Transfusion in Uganda." *British Red Cross Society Quarterly Review* 36 (1949): 106–9.
Moore, Jim. "The Puzzling Origins of AIDS." *American Scientist* 92, no. 6 (2004): 540–47.
Mottoulle, Léopold. "L'organisation du service médical et la situation sanitaire générale à l'Union minière du Haut-Katanga, fin 1929." *Annales de la Société belge de médecine tropicale* 11 (1931): 253–55.
Moyo, Dambisa. *Dead Aid: Why Aid Is Not Working and How There Is Another Way for Africa.* New York: Farrar, Straus and Giroux, 2009.
Musambachime, Mwelwa C. "The Impact of Rumor: The Case of the Banyama (Vampire Men) Scare in Northern Rhodesia, 1930–1964." *International Journal of African Historical Studies* 21, no. 2 (1988): 201–15.
Mwabu, Germano. "Health Care Reform in Kenya: A Review of the Process." *Health Policy* 32, nos. 1–3 (April–June 1995): 245–55.
Nahmias, A. J., J. Weiss, X. Yao, F. Lee, R. Kodsi, M. Schanfield, T. Matthews, et al. "Evidence for Human Infection with an HTLV III/LAV-like Virus in Central Africa, 1959." *Lancet* 1, no. 8492 (May 31, 1986): 1279–80.

Ndege, George O. *Health, State, and Society in Kenya.* Rochester: University of Rochester Press, 2001.

Ndinya-Achola, J. O., H. Nsanzumuhire, and G. B. A. Okelo. "Some Possible Infectious Hazards Due to Blood Transfusion in Nairobi." *East African Medical Journal* 57, no. 1 (1980): 55–59.

Neill, Deborah Joy. *Networks in Tropical Medicine: Internationalism, Colonialism, and the Rise of a Medical Specialty, 1890–1930.* Stanford: Stanford University Press, 2012.

Nelson, Joan M., ed. *Economic Crisis and Policy Choice: The Politics of Adjustment in the Third World.* Princeton: Princeton University Press, 1990.

N'tita, I., K. Mulanga, C. Dulat, D. Lusamba, T. Rehle, R. Korte, and H. Jäger. "Risk of Transfusion-Associated HIV Transmission in Kinshasa Zaire." *AIDS* 5, no. 4 (1991): 437–39.

Nwagbo, D., B. Nwaloby, C. Akpala, W. Chukudebelu, A. Ikeme, J. Okaro, B. Onah, V. Okeke, P. Egbuciem, and A. Ikeagu. "Establishing a Blood Bank at a Small Hospital, Anambra State, Nigeria." *International Journal of Gynecology and Obstetrics* 59, supp. 2 (1997): 135–39.

Oehlecker, F. "Erfahrungen aus 170 direkten Bluttransfusionen von Vene zu Vene." *Archiv für klinische Chirurgie* 116 (1921): 714–15.

Ogot, Bethwell A., and William Robert Ochieng'. *Decolonization and Independence in Kenya, 1940–93.* Athens: Ohio University Press, 1995.

Okpara, R. "Transmission of HIV through Blood Transfusion." *African Health* 14, no. 5 (July 1992): 15–17.

Olumwullah, Osaak. *Dis-ease in the Colonial State: Medicine, Society, and Social Change among the AbaNyole of Western Kenya.* Westport, CT: Greenwood, 2002.

Osamo, N. O., L. A. Okafor, and S. Enebe. "Blood Group Distribution in Nigerians in Relation to Foreign Nationalities: Haematological and Anthropological Implications." *East African Medical Journal* 59, no. 8 (1982): 546–49.

Ouary, Gaston, Yann Goez, and Jacques Linhard. *Notes sur la réanimation—transfusion.* Algiers: Service de santé des troupes coloniales, 1944.

Paice, Edward. *World War I: The African Front.* New York: Pegasus, 2010.

Paraskeuis, Dimitrios, et al. "Dating the Origin and Dispersal of Hepatitis B Virus Infection in Humans and Primates." *Hepatology* 57, no. 3 (2013): 908–16.

Parker, Alison, K. L. Muiruri, and J. K. Preston. "Hepatitis-Associated Antigen in Blood Donors in Kenya." *East African Medical Journal* 48, no. 9 (1971): 470–75.

Pelis, Kim. "Blood Clots: The Nineteenth-century Debate over the Substance and Means of Transfusion in Britain." *Annals of Science* 54, no. 4 (1997): 331–60.

———. "Taking Credit: The Canadian Army Medical Corps and the British Conversion to Blood Transfusion in WWI." *Journal of the History of Medicine and Allied Sciences* 56, no. 3 (2001): 238–77.

Pepin, Jacques. *The Origins of AIDS.* Cambridge: Cambridge University Press, 2011.

Piot, Peter, T. C. Quinn, H. Taelman, F. M. Feinsod, K. B. Minlangu, O. Wobin, N. Mbendi, et al. "Acquired Immunodeficiency Syndrome in a Heterosexual Population in Zaire." *Lancet* 2, no. 8394 (July 14, 1984): 65–69.

Pirie, James Hunter Harvey. "Blood Testing Preliminary to Transfusion, with a Note on the Group Distribution among S.A. Natives." *Medical Journal of South Africa* 16 (January 1921): 109–12.

Premier congrès africain de transfusion sanguine, 7–12 Mars 1977, Yamoussoukro, Côte d'Ivoire, actes du congrès. Paris: Institut INNIT, 1981.

Prost, André. *Services de santé en pays africain: Leur place dans des structures socio-économiques en voie de développement.* Paris: Masson, 1970.

Quinn, T., Jonathan M. Mann, J. W. Curran, and Peter Piot. "AIDS in Africa: An Epidemiological Paradigm." *Science* 234, no. 4779 (November 21, 1986): 955–63.

Reitman, Judith. *Bad Blood: Crisis in the American Red Cross.* New York: Kensington Books, 1996.

République du Sénégal. Ministère de l'économie, des finances et du plan, direction de la planification et direction de la prévision et de la statistique. *Tableau de bord annuel de la situation sociale au Sénégal, 1991.* Dakar: Ministère de l'économie, 1991.

Rist, Gilbert. *The History of Development: From Western Origins to Global Faith.* French ed., 1996; 3rd ed., translated by Patrick Camiller. London: Zed, 2009.

Rodhain, Jérôme. "Giovanni Trolli (24 juin 1876–8 février 1942)." *Bulletin l'Institut royal colonial belge* 17 (1946): 169–74.

Rono, Joseph Kipkemboi. "The Impact of the Structural Adjustment Programmes on Kenyan Society." *Journal of Social Development in Africa* 17, no. 1 (2002): 81–98.

Ronsée, C.-S. "Anémies malariennes des enfants et transfusions sanguines, avec observations sur les groupes sanguins des Bakongo." *Mémoires, Institut royal colonial belge, Section des sciences naturelles et médicales* 20, no. 2 (1952): 1–64.

———. "Sur les anémies malariennes des enfants et les transfusions sanguine." *Comptes rendus du congrès scientifique, Elisabethville, 13–19 août 1950,* vol. 5, *Travaux de la Commission de médecine humaine et vétérinaire.* Brussels: Comité spécial du Katanga, 1950.

Rous, Peyton, and J. R. Turner. "A Rapid and Simple Method of Testing Donors for Transfusion." *Journal of the American Medical Association* 64, no. 24 (1915): 1980–82.

Roux, Jean-Paul. *Le Sang: Mythes, symboles et réalités.* Paris: Fayard, 1988.

Ryder, R. W., H. C. Whittle, T. Wojiecowsky, W. M. Moffat, B. A. Baker, E. Sarr, and F. Oldfield. "Screening for Hepatitis B Virus Markers Is Not Justified in West African Transfusion Centres." *Lancet* 2, no. 8400 (August 25, 1984): 449–52.

Sankalé, M., H. Ruscher, and Y. Touré. "Accidents et incidents de la transfusion sanguine et leur prévention dans un service de médecine à Dakar." *Bulletin de la Société médicale d'Afrique noire de langue française* 18, no. 3 (1973): 307–11.

Scheyer, Stanley, and David Dunlop. "Health Services and Development in Uganda." In Dodge and Wiebe, *Crisis in Uganda,* 25–42.

Schindler, Elisabeth-Brigitte. *Le Centre de transfusion sanguine de la Croix-Rouge de Burundi, Son organisation et ses activités.* Bern: Croix-Rouge suisse, 1976.

Schneider, William H. "Blood Transfusion between the Wars." *Journal of the History of Medicine and Allied Sciences* 58, no. 2 (April 2003): 187–224.

———. "Blood Transfusion in Peace and War, 1900–1918." *Social History of Medicine* 10, no. 1 (April 1997): 105–26.

———. "The History of Research on Blood Group Genetics: Initial Discovery and Diffusion." *History and Philosophy of the Life Sciences* 18, no. 3 (1996): 273–303.

Schneider, William H., and Ernest Drucker, "Blood Transfusions in the Early Years of AIDS in Sub-Saharan Africa." *American Journal of Public Health* 96, no. 6 (June 2006): 984–94.

Schram, Ralph. *A History of the Nigerian Health Services.* Ibadan: Ibadan University Press, 1971.

Serafino, G., H. Tossou, and E. Goudote. "Réflexions sur cinq accidents transfusionnels par incompatibilité majeure." *Bulletin de la Société médicale d'Afrique noire de langue française* 6 (1961): 403–7.

Sharp, Paul M., and Beatrice H. Hahn. "Origins of HIV and the AIDS Pandemic." *Cold Spring Harbor Perspectives in Medicine* 1, no. 1 (September 2011): 1–22.

Sodahlon, Y. K., A. Y. Segbena, M. Prince-David, K. A. Gbo, M. P. Fargier, M. L. North, and D. J. Malvy. "Sécurité transfusionnelle dans un contexte de ressources limitées: Processus de mise en place de la politique nationale transfusionnelle au Togo." *Cahiers d'études et de recherches francophones / Santé* 14, no. 2 (April–June 2004): 115–20.

Spaander, J. "Le role de la Croix-Rouge néerlandaise dans le domaine de la transfusion sanguine." *Revue d'hématologie* 5, nos. 3–4 (1950): 497–507.

Spedener, D. "Le traitement des pneumonies des noirs par transfusion de sang des convalescents." *Bulletin médical du Katanga* 1 (1924): 234–38.

Sturgis, Cyrus C. "The History of Blood Transfusion." *Bulletin of the Medical Library Association* 30, no. 2 (January 1942): 105–12.

Sy, Baba. "Fonctionnement de la banque du sang de la Côte d'Ivoire." *Transfusion* (Paris) 3, no. 1 (1960): 47–51.

Tebit, Denis M., and Eric J. Arts. "Tracking a Century of Global Expansion and Evolution of HIV to Drive Understanding and to Combat Disease." *Lancet Infectious Diseases* 11 no. 1 (January 2011): 45–56.

Timberg, Craig, and Daniel Halperin. *Tinderbox: How the West Sparked the AIDS Epidemic and How the World Can Finally Overcome It.* New York: Penguin, 2012.

Titmuss, Richard M. *The Health Services of Tanganyika: A Report to the Government.* London: Pitman Medical Publishers, 1964.

Trolli, G. "Le service médical." In *L'essor économique belge: Expansion coloniale,* edited by Fernand Passelecq, 2 vols., 1:161–69. Brussels: L. Desmet-Verteneuil, 1932.

Trowell, Hubert Carey. *Non-infective Disease in Africa.* London: Edward Arnold, 1960.

United States. Department of Health and Human Services. U.S. Food and Drug Administration. "Keeping Blood Transfusions Safe: FDA's Multi-layered Protections for Donated Blood." Accessed August 8, 2010. http://www.fda.gov/BiologicsBloodVaccines/SafetyAvailability/BloodSafety/ucm095522.htm.

———. Department of State. Office of the U.S. Global AIDS Coordinator. "The President's Emergency Plan for AIDS Relief: Report on Blood Safety and HIV/AIDS." June 2006. Accessed September 2, 2011. http://www.state.gov/documents/organization/74125.pdf.

Valcke, George. "Note." *Annales de la Société belge de médecine tropicale* 14 (1934): 432–33.

van de Perre, P., D. Munyambuga, G. Zissis, J. P. Butzler, D. Nzaramba, N. Clumeck. "Antibody to HTLV-III in Blood Donors in Central Africa." *Lancet* 1, no. 8424 (February 9, 1985): 336–37.

van de Perre, P., D. Rouvroy, P. Lepage, J. Bogaerts, P. Kestelyn, J. Kayihigi, A. C. Hekker, J. P. Butzler, and N. Clumeck. "Acquired Immunodeficiency Syndrome in Rwanda." *Lancet* 2, no. 8394 (July 14, 1984): 62–65.

van de Walle, Nicolas. *African Economies and the Politics of Permanent Crisis, 1979–1999.* Cambridge: Cambridge University Press, 2001.

van Hulst, M., C. T. Smit Sibinga, and M. J. Postma. "Health Economics of Blood Transfusion Safety—Focus on Sub-Saharan Africa." *Biologicals* 38, no. 1 (January 2010): 53–58.

van Nitsen, R. "La Pneumonie chez le noir, Essei de traitement." *Bulletin médical du Kataga* 1 (1924): 239–42.

Vaughan, Meghan. *Curing Their Ills: Colonial Power and African Illness.* Stanford: Stanford University Press, 1991.

———. "Health and Hegemony: Representation of Disease and the Creation of the Colonial Subject in Nyasaland." In *Contesting Colonial Hegemony: State and Society in Africa and India,* edited by Dagmar Engels and Shula Marks, 173–201. London: British Academic Press, 1994.

Von Steffan, Evelyn. "Dr. Hantchef (1910–2002) in Memoriam." *Donor Recruitment International* 86 (November 2002): 12.

Vos, J., B. Gumodoka, J. Z. Ng'weshemi, F. C. Kigadye, W. M. Dolmans, and M. W. Borgdorff. "Are Some Blood Transfusions Avoidable? A Hospital Record Analysis in Mwanza Region, Tanzania." *Tropical and Geographical Medicine* 45, no. 6 (1993): 301–3.

Walter, O., and L. Langlo. "A Blood-Bank Service in a Rural Hospital in East Africa." *East African Medical Journal* 39 (December 1962): 702–7.

Watson-Williams, E. J., and P. K. Kataaha. "Revival of the Ugandan Blood Transfusion System 1989: An Example of International Cooperation." *Transfusion Science* 11, no. 2 (1990): 179–84.

Wertheim, Joel O., and Michael Worobey. "Dating the Age of the SIV Lineages That Gave Rise to HIV-1 and HIV-2." *PLoS Computational Biology* 5, no. 5 (2009): e1000377.

White, Luise. *Speaking with Vampires: Rumor and History in Colonial Africa.* Berkeley: University of California Press, 2000.

WHO and International Federation of Red Cross and Red Crescent Societies. *Towards 100% Voluntary Blood Donation: A Global Framework for Action.* Geneva: World Health Organization, 2010. Accessed November 12, 2011. http://www.who.int/bloodsafety/publications/9789241599696_eng.pdf.

Winsbury, Rex, ed. *Safe Blood in Developing Countries: The Lessons from Uganda.* Brussels: European Commission, 1995.

World Bank. *Accelerated Development in Sub-Saharan Africa: An Agenda for Action.* Washington, DC: World Bank, 1981.

Worobey, Michael, M. Gemmell, T. Haselborn, M. T. P. Gilbert, D. E. Teuwen, K. Kunstman, S. M. Wolinsky, et al. "Direct Evidence of Extensive Diversity of HIV-1 in Kinshasa by 1960." *Nature* 455, no. 7213 (October 2, 2008): 661–64.

Zhu, T., B. T. Korber, A. J. Nahmias, Edward Hooper, Paul M. Sharp, and D. D. Ho. "An African HIV-1 Sequence from 1959 and Implications for the Origin of the Epidemic." *Nature* 391, no. 6667 (February 5, 1998): 594–97.

Zucker, J. R., E. M. Lackritz, T. K. Ruebush, A. W. Hightower, J. E. Adungosi, J. B. Were, and C. C. Campbell. "Anaemia, Blood Transfusion Practices, HIV and

Mortality among Women of Reproductive Age in Western Kenya." *Transactions of the Royal Society for Tropical Medicine and Hygiene* 88, no. 2 (March–April 1994): 173–76.

Zuzarte, J. C., and E. G. Kasili. "Hepatitis B Antigen—A Review." *East African Medical Journal* 55, no. 8 (1978): 346–54.

INDEX